Inventing the Internet

Janet Abbate

The MIT Press
Cambridge, Massachusetts
London, England

First MIT Press paperback edition, 2000

© 1999 Massachusetts Institute of Technology

Set in New Baskerville by Wellington Graphics.
Printed and bound in the United States of America.

Library of Congress Cataloging-in-Publication Data

Abbate, Janet.
  Inventing the Internet / Janet Abbate.
    p.   cm. — (Inside technology)
  Includes bibliographical references and index.
  ISBN 0-262-01172-7 (hardcover : alk. paper), 0-262-51115-0 (pb)
  1. Internet (Computer network)—History. I. Title. II. Series.
TK5105.875.I57A23   1999
004.67′8′09—dc21   19-847647
                                                              CIP

Contents

Acknowledgments

Many people and institutions gave me support while I was working on this project. The Charles Babbage Institute helped fund my graduate studies through its Tomash Graduate Fellowship and gave me access to its collection of oral histories and archival materials on the Advanced Research Projects Agency. I am indebted to Arthur Norberg, Judy O'Neill, and the CBI's staff for paving the way for me and for and others interested in this important chapter of the history of computing. Brian Kahin and Jim Keller at Harvard shared their knowledge of Internet policy issues while I was a member of the Kennedy School of Government's Information Infrastructure Project. The IEEE History Center at Rutgers University provided a postdoctoral fellowship while I was writing the manuscript, and I benefited from stimulating conversations with Bill Aspray, Andy Goldstein, David Morton, Rik Nebeker, Mike Geselowitz, and other colleagues at Rutgers. I spent countless working hours at the Someday Café in Somerville and La Colombe in Philadelphia, each of which provided a congenial environment along with superb coffee.

I made use of a number of archival collections. In addition to the Charles Babbage Institute, I wish to thank Sheldon Hochheiser at the AT&T archives and the staffs of the ARPA and MIT archives. I am indebted to Martin Campbell-Kelly for collecting unpublished materials on networking activities at the National Physical Laboratory, and to Jon Agar at the National Archive for the History of Computing at the University of Manchester for assistance in using them. Jennie Connolly at the library of Bolt, Beranek and Newman (now a part of GTE Internetworking) was also extremely helpful. Members of the Internet community made my work immeasurably easier by documenting their own activities in online archives and by creating the information infrastructure that supports so much scholarly work today.

I would like to thank these participants in the story of the Internet who took the time to talk with me by telephone or in person: Donald Davies, Derek Barber, David Farber, Alex McKenzie, John Day, Stephen Lukasik, and Colonel Heidi Heiden. Many others, including the late Jon Postel, responded helpfully to email queries. I am grateful to Paul Baran, Donald Davies, Derek Barber, Leonard Kleinrock, Howard Frank, Alex McKenzie, Vint Cerf, and Barry Leiner for reading drafts of various chapters and offering insights and corrections.

Murray Murphey, Karin Calvert, and Carolyn Marvin helped me work out early versions of these ideas while I was a graduate student at the University of Pennsylvania. I also learned much from the responses of audiences at talks I gave at Syracuse University, the Massachusetts Institute of Technology, Drexel University, and the University of Pennsylvania and at Large Technical Systems conferences in Sweden and France. Thomas P. Hughes, advisor and friend, encouraged my first attempts to write a history of the ARPANET long before the Internet became a household word. His advice and support over the years have been invaluable.

A number of friends and colleagues generously read and discussed portions of the manuscript, including Arthur Norberg, Bill Aspray, Martin Campbell-Kelly, Atsushi Akera, Elijah Millgram, and Judith Silverman. Paul Ceruzzi offered useful advice and the opportunity to discuss my ideas with other historians of computing. Robert Morris was always willing to share his expertise and enthusiasm on network matters and helped me appreciate many technical issues. Susan Garfinkel challenged me to think more creatively about the Internet as a cultural phenomenon. Special thanks are due my friends and writers' workshop partners Amy Slayton and David Morton, who commented on many drafts and whose patience, insight, and humor helped pull me through when the project seemed overwhelming. Finally, my love and deepest thanks to my sisters and brothers—Edith, Kennan, Lauren, Alain, Alison, Matthew, and Geoff—and their families, who provided moral support and laughs over the years, and to my parents, Anne and Mario Abbate, who stood behind me unfailingly. I couldn't have done it without you.

Inventing the Internet

Introduction

Between the 1960s and the 1980s, computing technology underwent a dramatic transformation: the computer, originally conceived as an isolated calculating device, was reborn as a means of communication. Today we take it for granted that information can travel long distances instantaneously. For many Americans, and for an increasing portion of the world's population, it has become easy and commonplace to send electronic mail or to access online multimedia information. The transcendence of geographic distance has come to seem an inherent part of computer technology. But in the early 1960s, when computers were scarce, expensive, and cumbersome, using a computer for communication was almost unthinkable. Even the sharing of software or data among users of different computers could be a formidable challenge. Before the advent of computer networks, a person who wanted to transfer information between computers usually had to carry some physical storage medium, such as a reel of magnetic tape or a stack of punch cards, from one machine to the other. Modems had been introduced in the late 1950s, but setting up a telephone connection between two machines could be an expensive and error-prone undertaking, and incompatibilities between computers compounded the difficulty of establishing such communications. A scientist who needed to use a distant computer might find it easier to get on a plane and fly to the machine's location to use it in person.

The worldwide system called the Internet played a major role in developing and popularizing network technology, which placed computers at the center of a new communications medium. Between the late 1960s and the 1990s, the Internet grew from a single experimental network serving a dozen sites in the United States to a globe-spanning system linking millions of computers. It brought innovative data communications techniques into the mainstream of networking practice, and it enabled a large number of Americans to experience the possibilities of cyberspace for the first time. By making long-distance interaction among different types of computers a common-place reality, the Internet helped redefine the practice and the meaning of computing.

Like all technologies, the Internet is a product of its social environment. The Internet and its predecessor, the ARPANET, were created by the US Department of Defense's Advanced Research Projects Agency (ARPA), a small agency that has been deeply involved in the development of computer science in the United States. My curiosity about the Internet grew out of my experiences as a computer programmer in the mid 1980s, when few people outside the field of computer science had heard of this network. I was aware that the Internet had been built and funded by the Department of Defense, yet here I was using the system to chat with my friends and to swap recipes with strangers—rather like taking a tank for a joyride! This apparent contradiction goes to the heart of the Internet's history, for the system evolved through an unusual (and sometimes uneasy) alliance between military and civilian interests.

The history of the Internet holds a number of surprises and confounds some common assumptions. The Internet is not a recent phenomenon; it represents decades of development. The US military played a greater part in creating the system than many people realize, defining and promoting the Internet technology to serve its interests. Network projects and experts outside the United States also made significant contributions to the system that are rarely recognized. Above all, the very notion of what the Internet is—its structure, its uses, and its value—has changed radically over the course of its existence. The network was not originally to be a medium for interpersonal communication; it was intended to allow scientists to overcome the difficulties of running programs on remote computers. The current commercially run, communication-oriented Internet emerged only after a long process of technical, organizational, and political restructuring.

The cast of characters involved in creating the Internet goes far beyond a few well-known individuals, such as Vinton Cerf and Robert Kahn, who have been justly celebrated for designing the Internet architecture. A number of ARPA managers[1] contributed to the Internet's development, and military agencies other than ARPA were active in running the network at times. The manager of the ARPANET project, Lawrence Roberts, assembled a large team of computer scientists that included both accomplished veterans and eager graduate students, and he drew on the ideas of network experimenters in the United States and in the United Kingdom. Cerf and Kahn also enlisted the help of computer experts from England, France, and the United States when they decided to expand the ARPANET into a system of interconnected networks that would become known as the Internet. As the popularity of networking spread, a new set of interest groups—telecommunications carriers, vendors of network products, international standards bodies—exerted influence on the evolution of the Internet. The National Science Foundation took over responsibility for the Internet in the 1980s, when ARPA willingly gave it up, only to turn the network over to private businesses in the 1990s. And far from the American centers of networking, at the CERN physics laboratory in Geneva, Tim Berners-Lee took advantage of the Internet's unique capabilities to invent an application that he called the World Wide Web. These individuals and organizations had their own agendas, resources, and visions for the future of the Internet. The history of the Internet is not, therefore, a story of a few heroic inventors; it is a tale of collaboration and conflict among a remarkable variety of players.

In this book I trace the history of the Internet from the development of networking ideas and techniques in the early 1960s to the introduction of the World Wide Web in the 1990s. I have chosen to focus on a set of topics that illuminate what I believe to be the most important social and cultural factors shaping the Internet. In chapter 1, I present the development of packet switching, the main technique used in the Internet, as a case study of how technologies are socially constructed. In chapter 2, I describe the creation of the ARPANET and discuss the significance of ARPA's unique system-building strategies. In its initial form the ARPANET was little more than an experimental collection of hardware and software; in chapter 3, I recount the struggles of the ARPANET's early users to find some practical applications for this infrastructure and their eventual success with electronic mail. In chapter 4, I describe the unusual convergence of defense and research interests that resulted in the creation of the Internet and the overlooked role of the military in the transition from ARPANET to Internet technology. In chapter 5, I place the Internet in the context of contemporary networking efforts around the world, examining the ways in which technical standards can be used as social and political instruments. In chapter 6, I survey the complex events and interactions that transformed the Internet into a commercially based popular medium in the 1990s and the accompanying fragmentation of control among diverse communities of producers and users. I conclude that the emergence of new applications such as the World Wide Web continues the trend of informal, decentralized, user-driven development that characterized the Internet's earlier history.

In telling the story of the Internet, I also try to fill a gap in historical writing about computers. Much of the literature on the history of computing has focused on changes in hardware, on the achievements of individual inventors, or on the strategies of commercial firms or other institutions.[2] Relatively few authors have looked at the social shaping of computer communications.[3] There have been many social and cultural studies of computing in recent years, including compelling analyses of networking by Sherry Turkle (1995), Gene Rochlin (1997), and Philip Agre (1998a,b), but these works tend not to examine in detail the origins of computer technologies, focusing instead on how they are used once they exist. In this book I hope to cross the divide that exists between narratives of production and narratives of use. I demonstrate that the kinds of social dynamics that we associate with the use of networks also came into play during their creation, and that users are not necessarily just "consumers" of a technology but can take an active part in defining its features. Indeed, the culture of the Internet challenges the whole distinction between producers and users. I also try to provide some historical grounding for cultural studies of the Internet by documenting the events and decisions that created the conditions of possibility for the Internet's current status as a popular communication medium and the associated social experiments in cyberspace.

Is there something unique about the Internet's seemingly chaotic development? What, if anything, can the history of the Internet tell us about the nature of technology? Perhaps the fluid, decentralized structure of the Internet should be viewed as typical of late-twentieth-century technological systems, as it exemplifies both the increased complexity of many "high-tech" fields and new forms of organization that favor flexibility and collaboration among diverse interest groups. In computing, especially, systems and organizations have had to adapt to survive the relentless pace of technological change. The Internet also shares the protean character of communications media: since "information" (that infinitely malleable entity) is at the heart of the technology, media are particularly susceptible to being adapted for new purposes. Communications media often seem to dematerialize technology, presenting themselves to the user as systems that transmit ideas rather than electrons. The turbulent history of the Internet may be a reminder of the very real material considerations that lie behind this technology and of their economic and political consequences.

As I have already suggested, one of my aims in this book is to show how military concerns and goals were built into the Internet technology. My account of the origins of the network demonstrates that the design of both the ARPANET and the Internet favored military values, such as survivability, flexibility, and high performance, over commercial goals, such as low cost, simplicity, or consumer appeal. These values have, in turn, affected how the network has been managed and used. The Department of Defense's ability to command ample economic and technical resources for computing research during the Cold War was also a crucial factor in launching the Internet. At the same time, the group that designed and built ARPA's networks was dominated by academic scientists, who incorporated their own values of collegiality, decentralization of authority, and open exchange of information into the system. To highlight these social and cultural influences on its design, I compare the Internet with networking projects from other contexts and other countries, showing how the ARPA approach differed from alternative networking philosophies. The wider history of networking also reveals the reciprocal influences between the Internet and other projects in the United States and abroad.

I also emphasize the importance of network users in shaping the technology. In the early days of the ARPANET, the distinction between producers and users did not even exist, since ARPA's computer experts were building the system for their own use. Their dual role as users and producers led the ARPANET's builders to adopt a new paradigm for managing the evolution of the system: rather than centralize design authority in a small group of network managers, they deliberately created a system that allowed any user with the requisite skill and interest to propose a new feature. As access to the ARPANET and the Internet spread beyond the initial group of computer scientists, nonexpert users also exerted influence, improvising new ways of using the network and deciding which applications would become standard features of the system and which would fade away. I argue that much of the Internet's success can be attributed to its users' ability to shape the network to meet their own objectives. Electronic mail and the World

Wide Web are prominent examples of informally created applications that became popular, not as the result of some central agency's marketing plan, but through the spontaneous decisions of thousands of independent users.

In reconstructing the history of the Internet, I have been struck time and again by the unexpected twists and turns its development has taken. Often a well-laid plan was abandoned after a short time and replaced by a new approach from an unexpected quarter. Rapid advances, such as the introduction of personal computers and the invention of local-area networks, continually threatened to make existing network technologies obsolete. In addition, responsibility for operating the Internet changed hands several times over the course of its first thirty years or so. How, in the face of all this change and uncertainty, did the system survive and even flourish? I believe that the key to the Internet's success was a commitment to flexibility and diversity, both in technical design and in organizational culture. No one could predict the specific changes that would revolutionize the computing and communications industries at the end of the twentieth century. A network architecture designed to accommodate a variety of computing technologies, combined with an informal and inclusive management style, gave the Internet system the ability to adapt to an unpredictable environment.

The Internet's identity as a communication medium was not inherent in the technology; it was constructed through a series of social choices. The ingenuity of the system's builders and the practices of its users have proved just as crucial as computers and telephone circuits in defining the structure and purpose of the Internet. That is what the title of this book, *Inventing the Internet,* is meant to evoke: not an isolated act of invention, but rather the idea that the meaning of the Internet had to be invented—and constantly reinvented—at the same time as the technology itself. I hope that this perspective will prove useful to those of us, experts and users alike, who are even now engaged in reinventing the Internet.

1

White Heat and Cold War: The Origins and Meanings of Packet Switching

Of all the ARPANET's technical innovations, perhaps the most celebrated was packet switching. Packet switching was an experimental, even controversial method for transmitting data across a network. Its proponents claimed that it would increase the efficiency, reliability, and speed of data communications, but it was also quite complex to implement, and some communications experts argued that the technique would never work. Indeed, one reason the ARPANET became the focus of so much attention within the computer science community was that it represented the first large-scale demonstration of the feasibility of packet switching.[1] The successful use of packet switching in the ARPANET and in other early networks paved the way for the technique's widespread adoption, and at the end of the twentieth century packet switching continued to be the dominant networking practice. It had moved from the margins to the center, from experimental to "normal" technology.[2]

Many computer professionals have seen packet switching as having obvious technical advantages over alternative methods for transmitting data, and they have tended to treat its widespread adoption as a natural result of these advantages. In fact, however, the success of packet switching was not a sure thing, and for many years there was no consensus on what its defining characteristics were, what advantages it offered, or how it should be implemented—in part because computer scientists evaluated it in ideological as well as technical terms. Before packet switching could achieve legitimacy in the eyes of data communications practitioners, its proponents had to prove that it would work by building demonstration networks. The wide disparity in the outcomes of these early experiments with packet switching demonstrates that the concept could be realized in very different ways, and that, far from being a straightforward matter of a superior technology's winning out, the "success" of packet switching depended greatly on how it was interpreted.

Packet switching was invented independently by two computer researchers working in very different contexts: Paul Baran at the Rand Corporation in the United States and Donald Davies at the National Physical Laboratory in England. Baran was first to explore the idea, around 1960; Davies came up with his own version of packet switching a few years later and subsequently learned of Baran's prior work. Davies was instrumental in passing on the knowledge of packet switching that he and Baran had developed to Lawrence Roberts, who was in charge of creating the ARPANET. This chain of invention and dissemination has become a standard element of origin stories about the Internet; indeed, it is easy to get the impression that packet switching simply took a detour through the United Kingdom before re-emerging, unchanged, in the United States to fulfill its destiny as the underlying technology of the ARPANET.[3]

However, while Baran's and Davies's versions of packet switching had some basic technical similarities, their conceptions of what defined packet switching and of what it was good for were very different. Much of this difference was due to the strong political pressures that were brought to bear on computing research in the United Kingdom and in the United States. Large computer projects in both countries were developed in a context of government funding and control, and national leaders saw computers as a strategic technology for achieving important political goals. But in the very different policy contexts of the United States and the United Kingdom, packet switching took on different meanings for Baran, Davies, and Roberts. Packet switching was never adopted on the basis of purely technical criteria, but always because it fit into a broader socio-technical understanding of how data networks could and should be used.

Networking Dr. Strangelove: The Cold War Roots of Packet Switching in the United States

As the 1960s opened, relations between the United States and the Union of Soviet Socialist Republics were distinctly chilly. The USSR had launched its Sputnik satellite in 1957, setting off alarm in the United States over a "science gap" and prompting a surge of government investment in science and technology. A series of events kept the

Cold War in the public consciousness: an American U-2 spy plane was shot down over the USSR in 1960, the Berlin Wall went up in 1961, and 1962 brought the Cuban Missile Crisis. The shadow of nuclear war loomed over popular culture. The novels *On the Beach* (Shute 1957) and *Fail-Safe* (Burdick and Wheeler 1962)—both made into movies in the early 1960s—presented chilling accounts of nuclear war and its aftermath. And in 1964, movie theaters across the United States presented a brilliant black comedy of Cold War paranoia, *Dr. Strangelove* (Kubrick 1963).

Dr. Strangelove, though humorous, highlighted the vulnerability of the United States' communications channels to disruption by a Soviet attack, which might make them unavailable just when they were needed most. In the movie, a psychotic Air Force commander named Jack D. Ripper sets a nuclear holocaust in motion by invoking a strategy of mutual assured destruction called "Plan R." This plan—which allows Ripper to circumvent the president's authority to declare war—is specifically designed to compensate for a wartime failure in command, control, and communications. In the movie, an Air Force general explains:

Plan R is an emergency war plan in which a lower-echelon commander may order nuclear retaliation after a sneak attack—*if* the normal chain of command has been disrupted. . . . The idea was to discourage the Russkies from any hope that they could knock out Washington . . . as part of a general sneak attack and escape retaliation because of lack of proper command and control.

Plan R allows Ripper to launch a "retaliatory" attack even though no first strike has actually occurred. In reality (as the film's disclaimer states), the US Air Force never had any such strategy. Even before *Dr. Strangelove* opened, the Air Force was exploring a very different solution to the threat of a first strike: building a communications system that would be able to survive an attack and so that "proper command and control" could be maintained. As Edwards (1996, p. 133) has documented, Cold War defense analysts saw robust communications networks as a necessity in any nuclear confrontation: "Flexible-response strategy required that political leaders continue to communicate during an escalating nuclear exchange. . . . Therefore, preserving central command and control—political leadership, but also reconnaissance, data, and communications links—achieved the highest military priority." The need for "survivable communications" was generally recognized by the early 1960s. Among those intent on filling it was a researcher at the Air Force's premier "think tank," the Rand Corporation.

Founded by the Air Force in 1946 as an outgrowth of operations research efforts initiated during World War II, Rand (originally RAND, derived from "research and development") was a nonprofit corporation dedicated to research on military strategy and technology. Rand was primarily funded by contracts from the Air Force, though it served other government agencies as well. It attracted talented minds though a combination of high salaries, relative autonomy for researchers, and the chance to contribute to policy decisions of the highest importance (Baran 1990, pp. 10, 11). Edwards (1996, p. 116) notes that "Rand was the center of civilian intellectual involvement in defense problems of the 1950s, especially the overarching issue of nuclear politics and strategy." Rand's role was visible enough to be reflected in popular culture—for example, the fictional Dr. Strangelove turns to "the Bland Corporation" when he needs advice on nuclear strategy.[4] Because its approach to systems analysis emphasized quantitative models and simulation, Rand was also active in computer science research (Edwards 1996, pp. 122–124).

In 1959 a young engineer named Paul Baran joined Rand's computer science department. Immersed in a corporate culture focused on the Cold War, Baran soon developed an interest in survivable communications, which he felt would decrease the temptation of military leaders to launch a preemptive first strike:

Both the US and USSR were building hair-trigger nuclear ballistic missile systems. . . . If the strategic weapons command and control systems could be more survivable, then the country's retaliatory capability could better allow it to withstand an attack and still function; a more stable position. But this was not a wholly feasible concept, because long-distance communications networks at that time were extremely vulnerable and not able to survive attack. That was the issue. Here a most dangerous situation was created by the lack of a survivable communication system.   (Baran 1990, p. 11)[5]

Baran was able to explore this idea without an explicit contract from the Air Force (ibid., pp. 12, 16), since Rand had a considerable amount of open-ended funding that researchers could use to pursue projects they deemed relevant to the United States' defense concerns.[6]

Baran began in 1959 with a plan for a minimal communications system that could transmit a simple "Go/No go" message from the president to commanders by means of AM radio. When Baran presented this idea to military officers, they immediately insisted that they needed greater communications capacity. Baran spent the next three years formulating ideas for a new communications system that would combine survivability with high capacity (ibid., pp. 14–15). He envisioned a system would allow military personnel to carry on voice conversations or to use teletype, facsimile, or low-speed computer terminals under wartime conditions. The key to this new system was a technique that Baran (1960, p. 3) called "distributed communications." In a conventional communications system, such as the telephone network, switching is concentrated and hierarchical. Calls go first to a local office, then to a regional or national switching office if a connection beyond the local area is needed. Each user is connected to only one local office, and each local office serves a large number of users. Thus, destroying a single local office would cut off many users from the network. A distributed system would have many switching nodes, and many links attached to each node. The redundancy would make it harder to cut off service to users.

In Baran's proposed system, each of several hundred switching nodes would be connected to other nodes by as many as eight lines (figure 1.1). Several hundred multiplexing stations would provide an interface between the users and the network. Each multiplexing station would be connected to two or three switching nodes and to as many as 1024 users with data terminals or digital telephones. The switching was distributed among all the nodes in the network, so knocking out a few important centers would not disable the whole network. To make the system even more secure, Baran (1964a, volume VIII, section V) planned to locate the nodes far from population centers (which were considered military targets), and he designed the multiplexing stations with a wide margin of excess capacity (on the assumption that attacks would cause some equipment to fail). Baran added such military features as cryptography and a priority system that would allow high-level users to preempt messages from lower-level users.

To move data through the network, Baran adapted a technique known as "message switching" or "store-and-forward switching." A common example of message switching is the postal system. In a message switching system, each message (e.g., a letter) is labeled with its origin and its destination and is then passed from node to node through the network. A message is temporarily stored at each node (e.g., a post office) until it can be forwarded to the next node or the final destination. Each successive node uses the address information

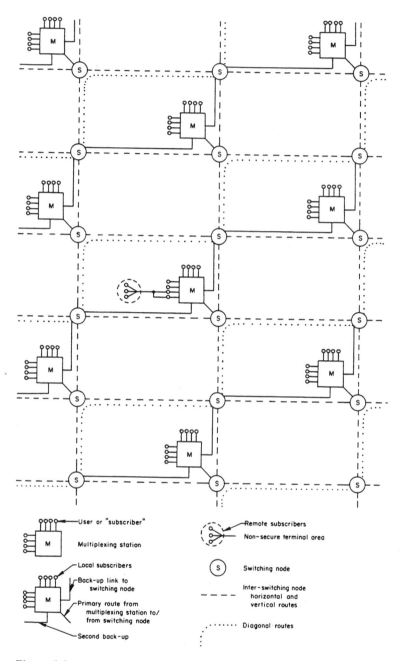

Figure 1.1
Paul Baran's design featuring highly connected switching nodes. Source:
Baran 1964a, volume VIII.

to determine the next step of the route. In the 1930s, message switching came into use in telegraphy: a message was stored on paper tape at each intermediate station before being transmitted to the next station. At first, telegraph messages were switched manually by the telegraph operators; however, in the 1960s telegraph offices began to use computers to store and route the messages (Campbell-Kelly 1988, p. 224).

For the postal and telegraph systems, message switching was more efficient than transmitting messages or letters directly from a source to a destination. Letters are stored temporarily at a post office so that a large number can be gathered for each delivery route. In telegraphy, message switching also addressed the uneven flow of traffic on the expensive long-distance lines. In periods of light traffic, excess capacity was wasted; when the lines were overloaded, there was a risk that some messages would be lost. Storing messages at intermediate stations made it possible to even out the flow: if a line was busy, messages could be stored at the switch until the line was free. In this way, message switching increased the efficiency, and hence the economy, of long-distance telegraphy.[7]

Besides appreciating the efficiency offered by message switching, Baran saw it as a way to make his system more survivable. Since the nodes in a message switching system act independently in processing the messages and there are no preset routes between nodes, the nodes can adapt to changing conditions by picking the route that is best at any moment. Baran (1964b, p. 8) described it this way: "There is no central control; only a simple local routing policy is performed at each node, yet the over-all system adapts." This increases the ability of the system to survive an attack, since the nodes can reroute messages around non-functioning parts of the network. Baran realized that survivability depended on more than just having redundant links; the nodes must be able to make use of those extra links. "Survivability," Baran wrote (1964a, volume V, section I), "is a function of switching flexibility." Therefore, his network design was characterized by distributed routing as well as distributed links.

Departures from Other Contemporary Systems
Paul Baran was not the first to propose either message switching or survivable communications to the military. Systems of both types already existed or were in development. A look at the state of the art in these areas makes it easier to see what aspects of Baran's ideas were

really innovative and why he saw opportunities to depart from contemporary practice in certain areas.

Message switching systems were nothing new to the Department of Defense, but the existing systems were cumbersome and inefficient. Baran discovered this when he served as a member of a Department of Defense committee charged with examining several existing or proposed store-and-forward data systems in the early 1960s. These systems had such low capacity that backlogs of messages tended to build up at the switches. Therefore, the switches had to be built with large storage capacity to hold the messages that were waiting to be forwarded, and the switching computers ended up being large and complex. Baran was convinced that a network could and should be built using much higher transmission speeds, eliminating the bottlenecks at the nodes. Besides the obvious benefit of getting messages delivered faster, a high-speed, low-storage system could have switching nodes that were much simpler and cheaper than those used in contemporary store-and-forward data systems. As Baran (1964b, p. 6) pointed out, although the high-speed system would be store-and-forward in its design, in practice messages would spend little time being stored at the nodes; to the user, therefore, a connection would seem to be real-time. Baran's argument (1990, p. 24) that it was possible to build a message switching network with fast end-to-end transmission of messages and small, inexpensive switches was a radical challenge to the existing understanding of such systems.

The concept of "distributed communications" (or "distributed networks") also predated Baran; indeed, his publications cite examples of the idea from mathematics, artificial intelligence, and civilian and military communications (Baran 1964a, volume V, section I). In particular, military planners had already proposed a variety of systems based on a network of decentralized nodes linked by multiple connections (ibid., section IV). Though they shared the idea of distributed communications, however, these other systems differed in essential ways from Baran's proposal. In particular, most of them seem to have entailed the use of simple broadcast techniques, with every message going to every destination, whereas Baran's system would route messages individually through the network.[8]

Most of the distributed systems Baran described were only proposals, not working systems. However, there was one large distributed communications network under actual development in the early 1960s. This was AUTOVON, designed and operated for the Department of Defense by the American Telephone and Telegraph Corporation. In 1961 AT&T had provided the Army with a communications network called the Switched Circuit Automatic Network, and in 1963 the corporation provided a similar network for the Air Force called North American Air Defense Command/Automatic Dial Switching. The Defense Communications Agency, which was charged with coordinating the provision of communications services throughout the armed services, decided to integrate these networks into a new system called the Continental United States Automatic Voice Network (CONUS AUTOVON). AUTOVON was not a message switching system; it was a special military voice network built on top of the existing civilian telephone network. It went into service with ten switching nodes in April of 1964 (Schindler 1982, pp. 266–269).

Describing the AUTOVON system, AT&T's magazine *Long Lines* (1965, p. 3) noted: "The top requirement is that the system can survive disaster." Survivability was sought in part by placing the switching centers in "hardened" sites, often underground, away from major metropolitan targets. The main survivability feature, however, was that the network was arranged in what AT&T called a "polygrid," with each switch connected to several links and with the links distributed evenly throughout the system (rather than having all connections routed through a few central switches). AT&T's publicity stressed that this redundant, decentralized system represented a sharp departure from the hierarchical structure used in the ordinary toll network. AUTO-VON had one node for every few hundred lines, whereas in the regular toll system a node typically served a few thousand lines. "The polygrid network," according to the system's architects, "plays a major role in the survivability of AUTOVON. Along with its other virtues of flexibility and economy, polygrid represents the best method that technology can now offer for the rapid and reliable connection of defense communications." (Gorgas 1968, p. 227)

Baran's approach differed from AT&T's in two significant ways. First, although AUTOVON had nodes distributed throughout the system, control of those nodes was concentrated in a single operations center, where operators monitored warning lights, analyzed traffic levels, and controlled system operations. If traffic had to be rerouted, it was done manually: operators at the control center would make the decision and then contact the operators at the switching nodes with instructions to change routes (Gorgas 1968, p. 223; *Long Lines* 1969). In Baran's network, control was fully distributed, as noted above.

Nodes would be individually responsible for determining routes, and would do so automatically without human intervention: "The intelligence required to switch signals to surviving links is at the link nodes and *not* at one or a few centralized switching centers." (Baran 1960, p. 3) Clearly such a system would be more survivable than one dependent on a single operations center—which, Baran noted, "forms a single, very attractive target in the thermonuclear era" (1964a, volume V, section II).

One implication of Baran's design was that the nodes would have to have enough "intelligence" to perform their own routing—they would have to be computers, not just telephone switches. This brings us to Baran's second departure from the AT&T approach: Baran envisioned an all-digital network, with computerized switches and digital transmission across the links. The complexity of routing messages would require computers at the nodes, since the switches would have to be able to determine, on their own, the best path to any destination, and to update that information as network conditions changed. Such computerized switches had never been designed before. "These problems," Baran acknowledged (1964b, p. 6), "place difficult requirements on the switching. However, the development of digital computer technology has advanced so rapidly that it now appears possible to satisfy these requirements by a moderate amount of digital equipment." Preserving the clarity of the signal would require that transmission be digital as well. One consequence of having a distributed network was that a connection between any two endpoints would typically be made up of many short links rather than a few long ones, with messages passing through many nodes on the way to their destinations. Having many links in a route was problematic for the transmission of ordinary analog signals: the signal degenerated slightly whenever it was switched from one link to another, and distortion accumulated with each additional link. Digital signals, on the other hand, could be regenerated at each switch; thus, digital transmission would allow the use of many links without cumulative distortion and errors. Digital transmission was still a novelty at the time; Bell Labs had only begun developing its T1 digital trunk lines in 1955, and they would not be ready for commercial service in the Bell System until 1962 (O'Neill 1985).[9]

Baran's system would push contemporary switching and transmission technology to their limits, so it is understandable that contemporary experts reacted skeptically to his claims. The engineers in AT&T's

Long Lines Division, which ran the long-distance telephone service and the AUTOVON system, tended to be familiar only with analog technology, and they doubted Baran's claims that an all-digital system could transcend the well-known limits on the number of links per call (Baran 1990, p. 18).[10] Whereas in AUTOVON there was a maximum of seven links in any route, Baran's simulations of network routing in a small version of his system showed as many as 23 links between endpoints (Gorgas 1968, 223; Baran 1964b, p. 7, figure 11). Evidently, Baran's position outside the community of analog communications practitioners and his awareness of the potential of computer techniques made it easier for him to question the accepted limits. He had no stake in analog telephony, and his training and background in computing made it easier for him to envision an all-digital system as a way of achieving his goal of distributed communication.

And Baran's system departed from traditional telephone company practice in other ways that show the effect of Cold War military considerations on his design assumptions. For instance, AT&T tried to increase the reliability of the phone system as a whole by making each component as reliable as possible, and for an additional fee would provide lines that were specially conditioned to have lower error rates. Baran chose instead to make do with lower-quality communications links and to provide redundant components to compensate for failures. Conditioned lines would be too expensive for a system with so many links, and in any case the reliability of individual components could not be counted on in wartime conditions. "Reliability and raw error rates are secondary," observed Baran (1964b, pp. 4–5). "The network must be built with the expectation of heavy damage anyway."

Packet Switching in Baran's System

Baran's proposed network began as a distributed message switching system. His final innovation was to alter message switching to create a new technique: packet switching. In his system a message could be anything from digitized speech to computer data, but the fact that these messages were all sent in digital form—as a series of binary numbers ("bits")—meant that the information could be manipulated in new ways. Baran proposed that, rather than sending messages of varying sizes across the network, messages should be divided into fixed-size units that he called "message blocks." The multiplexing stations that connected users to the network would be responsible for dividing outgoing messages into uniform blocks of 1024 bits. A short message could be sent as a single block; longer messages would require multiple message blocks. The multiplexer would add to each block a header specifying the addresses of the sending and receiving parties as well as other control information. The switching nodes would use the header information to determine what route each block should take to its destination; since each block was routed independently, the different blocks that made up a single message might be sent on different routes. When the blocks reached their destination, the local multiplexer would strip the header information from each block and reassemble the blocks to form the complete message. This idea would eventually be widely adopted for use in computer networks; the message blocks would come to be called "packets" and the technique "packet switching."[11]

For all its eventual significance, the decision to transmit data as packets was not the original focus of Baran's work. As the title of his eleven-volume work *On Distributed Communications* indicates, Baran began with the idea of building a distributed network—an idea that had already been identified with survivability by people working in military communications (Baran 1964a, volume V). In describing the system, Baran tended to stress the idea of link redundancy, rather than other elements such as packet switching.[12] But as he developed the details of the system, the use of message blocks emerged as a fundamental element. By the time he wrote the final volume of the series, Baran had changed the name he used to refer to the system to reflect the new emphasis: "While preparing the draft of this concluding number, it became evident that a distinct and specific system was being described, which we have now chosen to call the 'Distributed Adaptive Message Block Network,' in order to distinguish it from the growing set of other distributed networks and systems." (Baran 1964a, volume XI, section I) What, then, was so important about packet switching? What did it mean to Baran and his sponsors?

Transmitting packets rather than complete messages imposed certain costs on the system. The interface computers had to perform the work of dividing users' outgoing messages into packets and of reassembling incoming packets into messages. There was also the overhead of having to include address and control information with each packet (rather than once per message), which increased the amount of data that had to be transmitted over the network. And since packets from a single message could take different routes to their destination, they might arrive out of sequence, which meant that there had to be provisions for reassembling them in the proper order. All this made the system more complex and presented more opportunities for failure. For Baran, these costs were outweighed by his belief that packet switching would support some of the fundamental goals of the system.

Packet switching offered a variety of benefits. Baran was determined to use small, inexpensive computers for his system, rather than the huge ones he had seen in other message switching systems, and he was aware that the switching computers would have to be simple in order to be both fast and inexpensive. The use of fixed-size packets rather than variable-size messages could simplify the design of the switching node. Another advantage for the military was that breaking messages into packets and sending them along different routes to their destination would make it harder for spies to eavesdrop on a conversation. But the biggest potential reward was efficient and flexible transmission of data. "Most importantly," wrote Baran (1964b, p. 6), "standardized data blocks permit many simultaneous users, each with widely different bandwidth requirements[,] to economically share a broad-band network made up of varied data rate links." In other words, packet switching allowed a more efficient form of multiplexing (sharing of a single communication channel by many users).

In the conventional telecommunications systems of the early 1960s, the usual form of multiplexing was by frequency division: each caller would be assigned a particular frequency band for their exclusive use for the duration of their connection. If the caller did not talk or send data continuously, the idle time would be wasted. In an alternative method, called "time division multiplexing," time is divided into short intervals, and each user in turn is given a chance to transmit data for the duration of one interval. Only users with data to transmit are offered time slots, so no slots go idle as long as anyone has data to transmit; this makes time division multiplexing more efficient for usage situations where bursts of information alternate with idle periods. Since computer data tends to have this "bursty" characteristic, Baran (1964b, p. 6) felt that time division was a more "natural" form of multiplexing for data transmission. And since the time slot accommodated a fixed amount of data, Baran believed that the use of fixed-size message blocks was a prerequisite for time division multiplexing. Thus, he associated packet switching with time division multiplexing and its promise of efficient data transmission.[13]

Packet switching would also make it easier to combine links having different data rates in the network. The data rate is the number of bits per second that can be transmitted on a given link. In the conventional telephone system, each caller is connected at a fixed data rate, and data must flow into and out of a switch at predetermined rates. With packet switching, data flowing into a switch can be divided among the outgoing links in a variety of ways, rather than having to be sent out at a fixed rate. This would make it easier for devices transmitting data at different rates (computers and digital telephones, for instance) to share a link to the network. The system could also take advantage of new media, such as low-cost microwave transmission, that had different data rates than the standard phone company circuits. Though packet switching made the system more complex in some respects, in other ways it made the system simpler and less costly to build.

In sum, packet switching appealed to Baran because it seemed to meet the requirements of a survivable military system. Cheaper nodes and links made it economically feasible to build a highly redundant (and therefore robust) network. Efficient transmission made it possible for commanders to have the higher communications capacity they wanted. Dividing messages into packets increased security by making it harder to intercept intelligible messages. Packet switching, as Baran understood it, made perfect sense in the Cold War context of his proposed system.

The Impact of Baran's Work

For a brief time after its publication in 1964, it seemed that Baran's *On Distributed Communications* might soon become the blueprint for a nationwide distributed packet switching network. In August of 1965, Rand officially recommended that the Air Force proceed with research and development on a "distributed adaptive message-block network." Enthusiastic about the proposal, Air Force representatives sent it for review to the Defense Communications Agency, which oversaw the provision of military communications services (Baran 1990, Attachment 2). The DCA was one of many agencies that had been created in an attempt to bring military operations under the central control of the Department of Defense rather than allowing each of the armed services to build its own systems.[14] In accordance with this centralizing strategy, the DoD administration made it clear during the review process that any new network would be built not by Air Force contractors but by the DCA, which had no expertise in digital technology. Baran and his Air Force sponsors, doubting that the DCA would be able to build the system that Baran had described, reluctantly decided to scrap the proposal rather than risk having it executed badly, which would waste large sums of money and perhaps discredit Baran's ideas (Baran 1990, pp. 33–35).[15]

Though the proposed network was never built, Baran's ideas were widely disseminated among researchers interested in new communications technologies. Following Rand's standard practice, Baran presented his work to various outside experts for comment as he was developing his ideas.[16] Eleven volumes of reports published in 1964 were widely distributed to individuals, government agencies, Rand depository libraries, and other people working in the field. The first volume was also published as an article in the March 1964 issue of *IEEE Transactions on Communications Systems,* and an abstract appeared in the August 1964 issue of *IEEE Spectrum* (a magazine for electrical and computing engineers with an estimated circulation of 160,000).[17] Baran also lectured on his work at various universities (Baran 1990, pp. 32–33, 36). It is not clear how many researchers were immediately influenced by Baran's ideas through these channels. Most academic computer scientists were not concerned with the survivability of communications, and they may not have seen the applicability of Baran's research to their own interests. Several years later, however, his work would begin to receive wide attention as one of the technical foundations of the ARPANET. Curiously enough, the connection between these two American networking efforts would be made via a laboratory in England.

Forging Packet Switching in the White Heat: Networks and Nationalism in the United Kingdom

In the early 1960s, while the United States was caught up in the Cold War, the United Kingdom was experiencing political upheaval of a different type. Just as the Americans were worried about a "science gap" between their country and the USSR, so there were widespread fears in the United Kingdom of a "technology gap" with the United States. Harold Wilson was elected leader of the British Labour Party in 1963, at a time when that party, and much of the general population, felt that the UK was facing an economic crisis. Politicians on all sides warned that the UK was falling behind the other industrial powers in its exploitation of new technologies, that there was a "brain drain" of British scientists to other countries, and that the country's technological backwardness was at least partly responsible for its economic malaise (Coopey and Clarke 1995; Edgerton 1996, pp. 53, 57).

Wilson addressed the technology issue head on in a speech to the Labour Party's annual conference at Scarborough on 2 October 1963. Calling on labor and management to join in revitalizing British industry, Wilson stressed the importance of keeping up with the ongoing scientific and technological revolution, and he invoked a stirring vision of a new United Kingdom "forged in the white heat of this revolution" (quoted in Edgerton 1996, p. 56). The speech created a sensation in the British media, and Wilson was praised in newspapers across the political spectrum for capturing the concerns of the times and remaking Labour's supposedly anti-progress image.[18] When Labour came to power in the 1964 general election, Wilson was eager to act on his vision by implementing a new economic and technological regime for the United Kingdom.

Wilson's plans included reversing the "brain drain" by training more scientists and giving them the status and the facilities that would persuade them to stay in the United Kingdom, by rationalizing existing industries and creating new high-tech industries, and by shifting resources from unproductive defense and "prestige" areas (such as aerospace and nuclear energy) to commercial applications. To oversee national technological development, Wilson created the Ministry of Technology, a major new department that assumed control of the Atomic Energy Authority, the Ministry of Aviation, the National Research Development Corporation, and a number of national laboratories (Edgerton 1996, pp. 65–70). Mintech, as it came to be called, had two main aims: to transfer the results of scientific research to industrial development, and to intervene in industry so as to make private enterprise more efficient and competitive. Mintech was to have, in Wilson's words (1971, p. 8), "a very direct responsibility for increasing productivity and efficiency, particularly within those industries in urgent need of restructuring or modernisation." These industries included machines tools, aviation, electronics, shipbuilding, and—above all—computing.

Wilson feared that the British computer industry would be destroyed by competition from the United States unless the government intervened quickly. He later recalled: "When, on the evening we took office, I asked Frank Cousins to become the first Minister of Technology, I told him that he had, in my view, about a month to save

the British computer industry and that this must be his first priority."
(Wilson 1971, p. 9) Cousins responded by increasing funding for
National Research Development Corporation, which gave development funds to corporations that wanted to commercialize government
research, and by using government contracts to encourage the introduction of new computer products (ibid., p. 63). In addition, Mintech
and the Industrial Reorganization Corporation were responsible for
pushing British corporate mergers to create large companies, such as
International Computers Limited, which would supposedly have the
critical mass of resources to compete internationally (Hendry 1990,
pp. 155–157; Wilson 1971, p. 63). In 1965 Mintech also took over a
government initiative called the Advanced Computer Techniques Project, which had been set up in 1960 to help spin off government-
sponsored computing research to industry. Under Wilson, computing
research was expected to serve economic aims, and the possibility of
government intervention was always present.

One of the British scientists who took the lead in computing research
was Donald W. Davies of the National Physical Laboratory in Teddington, a suburb of London. The NPL—established in 1899 to
determine values for physical constants, to standardize instruments for
physical measurements, and to perform similar activities involving
standards and materials testing (Pyatt 1983, pp. 157–158)—had first
become involved in computing in 1946, when a team at the laboratory,
following a proposal by Alan Turing, built an early stored-program
digital computer called the Pilot ACE. Davies had joined the NPL in
1947 and had worked on the Pilot ACE; in 1960 he had become
superintendent of the division in charge of computing science, and in
1965 he had been named technical manager of the Advanced Computer Techniques Project (Campbell-Kelly 1988, pp. 222–223).
Davies's position kept him in touch with the latest advances in computing technology and with the government's plans to use that technology to aid the British economy.

If the watchword for Baran was survivability, the priority for Davies
was interactive computing. Davies was one of many researchers who
hoped to improve the user friendliness of computers. Computers of
the early 1960s were expensive and in high demand. This meant that
their operating systems were designed for maximum efficiency in the
use of the computer's central processor. To achieve this, the typical
operating system of the early 1960s used batch processing, a technique
in which a number of computer programs would be collected and

loaded into the computer together to be executed in succession. Running programs in batches was efficient because it minimized the time the computer spent idle, waiting for data to be loaded or unloaded. The disadvantage of batch processing was that it did not allow users direct interaction with or an immediate response from the computer. As a result, computer users often experienced batch processing as slow, difficult, and tedious.

In the typical programming cycle, the user of a batch processing computer would begin by writing out a program on paper. Then the user or a keypunch operator would punch holes in a set of computer cards to represent the written instructions. The user would bring the deck of punched cards to the computer center, where an operator would feed them into a punched-card reader and transfer the data to magnetic tape. When the computer became available, the operator would load the tape and run its batch of programs, and eventually he or she would return a printout of the results to the various programmers. If a user's program turned out to have errors, the user would have to rewrite it, punch another set of cards, and submit the cards again, perhaps waiting hours for a chance to rerun the program and collect the results. Often users had to repeat this cycle numerous times before a program would work correctly.

Batch processing rationalized the flow of input to the computer, but it was frustrating and inefficient for the programmer. In the late 1950s computer scientists began to talk about a possible alternative, which they called "time sharing." Instead of running a single program from start to finish before going on to the next one, a time sharing operating system would cycle between a number of programs, devoting a fraction of a second of processing time to each one before going on the next (figure 1.2). The wait between cycles would be so short that users would have the impression of continuous interaction with the machine, just as moviegoers have the impression of seeing continuous motion on the screen rather than a rapid succession of still frames.

By sharing the computer's processor among multiple users, time sharing addressed the mismatch between the pace of human action and the much faster processing of the computer. When a computer serves a user at a interactive terminal, it spends most of its time waiting for commands; very little time is spent actually processing data. If a computer can serve many terminals at once, it will spend less time idle and more time doing productive work, which increases the efficiency—and therefore the economically feasibility—of interactive computing.

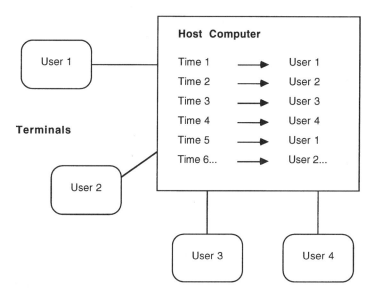

Figure 1.2
Time sharing.

Though time sharing is not necessarily synonymous with interactive computing,[19] the two ideas became closely associated. Time sharing was seen by its proponents as the innovation that would liberate computer users from their punched cards and allow direct and easy interaction with the machine.

The first proposals for time sharing operating systems were presented independently in 1959 by Christopher Strachey of the National Research Development Corporation in the United Kingdom and John McCarthy of the Massachusetts Institute of Technology in the United States. Time sharing caused tremendous excitement in the field of computing, and both academic researchers and industry analysts predicted that it would be the wave of the future. By the mid 1960s, research centers in the United Kingdom and in the United States were using time sharing computers regularly, and computer manufacturers were rushing to bring time sharing products to the market. Businesses began to spring up that offered access to time sharing machines on a commercial basis to customers who would rent or buy a terminal, connect to the service using a modem and a telephone line, and access the service's computers for an hourly rate. Many people thought that time sharing represented the future of interactive computing, since few if any anticipated the advent of small, inexpensive "personal" computers in the late 1970s.

Davies became interested in time sharing during a 1965 trip to the United States. He had gone there to participate in a computing conference being held at MIT, and he took the opportunity to visit several American computing research sites (Davies 1986, pp. 4–5). Both the conference and his site visits made it clear to him that interest in and knowledge about time sharing were much more widespread in the United States than in the United Kingdom. When Davies returned to the National Physical Laboratory, he decided to organize a seminar on time sharing to disseminate these ideas to the British computing community. The seminar was held in November of 1965, and a number of British and American researchers were invited.

It was during these discussions that Davies became aware of a widely perceived obstacle to interactive computing: inadequate data communications. In early time sharing systems, the terminals had been directly connected to the computer and were located in an adjacent terminal room. As time went on, people began locating terminals at some distance from the computer itself, either for the user's own convenience or, in the case of commercial time sharing services, to offer access to customers over a wide geographic area. Distant terminals could be connected to the computer using dial-up[20] telephone links and modems, but long-distance telephone connections were very expensive, and for data transmission they were also inefficient. Computer messages, as noted earlier, tend to come in bursts with long pauses in between, so computer users paid dearly for telephone connections that were idle much of the time. The high cost of communications put pressure on users to work quickly, sacrificing the user friendliness for which time sharing had been invented.[21] Davies had a long-standing interest in switching techniques. As he thought about the data communications problem, he came up with the idea that a new approach to switching might offer a solution (Davies 1986, pp. 6–7). He knew that message switching was used in the telegraph system to make efficient use of lines, and he believed that by adapting this technique to computer communications he could achieve similar economies. Like Baran, Davies came from a background in computing, rather than communications, so he felt free to suggest a technique that departed from traditional communications techniques but took advantage of advances in computer technology. Davies proposed dividing messages into standard-size "packets" and having a network of computerized switching nodes that would use information carried in packet headers to route the packets to and from time sharing computers. He called this technique "packet switching."

Packet Switching in Davies's System
Like Paul Baran, Donald Davies saw that packet switching would allow many users to share a communication link efficiently. But Davies wanted that efficiency for a different purpose. Packet switching, in his view, would be the communications equivalent of time sharing: it would maximize access to a scarce resource in order to provide affordable interactive computing.[22]

In March of 1966, Davies presented his network ideas publicly for the first time, to an enthusiastic audience of people active in computing, telecommunications, and the military. Afterward, a man from the British Ministry of Defence gave Davies the surprising news that packet switching had already been invented a few years earlier by an American (Baran). The fact that the military man knew about this earlier development when Davies did not underscores the very different contexts in which packet switching evolved in the two countries. Baran's foremost concern had been survivability, which was underlined by his use of terms like "raid," "salvos," "target," "attack level," and "probability of kill" in describing the hostile conditions under which his system was expected to operate (Baran 1964b, p. 2). Davies, on the other hand, did not view packet switching as a way to make the network survivable; after reading Baran, he commented that "the highly connected networks there considered" were "not needed in a civil environment" (Davies 1966b, p. 21).[23] Davies thought the pressing need was for a network that could serve the users of commercial time sharing services. This assumption is evident in his plan to survey businesses' data communications requirements (Davies 1968a). It also shows up in Davies's efforts to make the system easy to use. In his proposal for a national network, he wrote: "A further aim requirement we must keep in mind constantly is to make the use of the system simple for simple jobs. Even though there is a communication system and a computer operating system the user must be able to ignore the complexities." (Davies 1966a, p. 2)

Packet switching served the aim of building a commercial system mainly by bringing down the cost of data communications. However, Davies found further meanings in packet switching that derived from his vision of a commercial system. One of the merits he saw in packet switching was that it helped achieve fairness in access to the network. In an ordinary message switching system, each message had to be sent in its entirety before the next message could begin. In a packet switching system, time division multiplexing would allow users to take turns transmitting portions of their messages. If a user had a short message, such as a single command for a time sharing system, the whole message could be sent in the first packet, while longer messages would take several time slots to transmit. This way, the user with the short message would not have to wait behind users with long messages (Campbell-Kelly 1988, p. 226). This kind of fairness was appropriate for a system where computers were serving the everyday needs of civilians, rather than transmitting life-or-death messages through a command hierarchy.

Ultimately, Davies thought, packet switching technology could become a commercial product that would contribute directly to Harold Wilson's plan to revitalize the British economy. In a 1965 proposal to have the General Post Office build a prototype for a national packet switching network, Davies (1965, p. 8) wrote:

Such an experiment at an early stage is needed to develop the knowledge of these systems in the GPO and the British computer and communications industry. . . . It is very important not to find ourselves forced to buy computers and software for these systems from [the] USA. We could, by starting early enough, develop export markets.[24]

Davies (1968c, p. 7) reiterated the need to compete with the United States in 1968, when he compared a proposed Mintech network with the planned ARPANET:

The proposal resembles the ARPA network being planned. . . . The sponsors of that project believe it will "spearhead" a new kind of data communication system to be developed on a nation-wide scale.
  A Mintech network would go beyond the present ARPA plans by providing for a variety of terminals as well as computer to computer communication. To be useful as a "spearhead" project it would need to be started soon and planned with as short a time scale as possible, coming into operation well before a national network.

For Davies, the network was not only a communications tool; it was also a way for British researchers to apply the "white heat" of scientific innovation to counteracting American dominance in the computer market.

Davies's concern with economics and user friendliness underscores the national context in which he conceived the idea of a packet switching network. Davies did not envision a world in which his proposed network would be the only surviving communications system. Rather, he saw a world in which packet switching networks would compete with other communications systems to attract and serve the business user and in which the United Kingdom would need to compete with the United States and other countries to offer innovative computer products.

In December of 1965, Davies proposed the idea of a national packet switching network that would provide inexpensive data communications across the United Kingdom (figure 1.3). He envisioned the network as offering a number of services to business and recreational users, including remote data processing, point-of-sale transactions, database queries, remote control of machines, and even online betting (Davies 1965). In his scheme, a backbone of high-capacity telephone lines would link major cities in the United Kingdom; the proposed network had multiple connections to most nodes, although it was not nearly as redundant as Baran's system.[25] Like Baran, the National Physical Laboratory group designed a network with a dynamic, distributed routing system, each node making routing decisions independently according to current conditions in the network. The nodes would be connected by high-speed telephone lines so as to provide fast response for interactive computing. Users would attach their computers, terminals, and printers to the nodes through dedicated interface computers at local sites.

Davies was convinced that a data communications infrastructure of the sort he was proposing would be necessary to keep the United Kingdom competitive in the information age, and he did not doubt that such a network would someday be built. However, the NPL did not have the resources or the authority to build such a large network on its own. This authority belonged to the General Post Office, which ran the national postal and telephone networks, but managers there had little knowledge of or interest in data communications. Since Davies felt there was no hope of convincing the GPO to collaborate on a national network, he decided that a small in-house experiment would be the only feasible alternative. In the summer of 1966 he made a second, much more modest proposal to build a prototype network at the NPL. This network, named "Mark I," would serve as a demonstration of packet switching, advance the state of knowledge in the field, and support the operational computing needs of the NPL's scientific and administrative personnel.

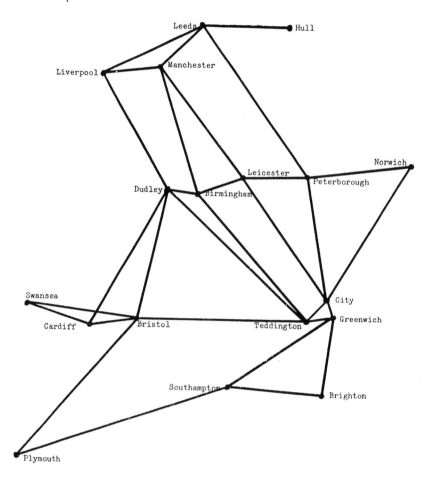

Figure 1.3
Donald Davies's proposed network for the United Kingdom (from archival copy).

The Mark I project started in 1967 with a development team headed by Derek Barber. Roger Scantlebury was the technical leader, Keith Bartlett oversaw hardware development, and Peter Wilkinson was in charge of software development (Campbell-Kelly 1988, pp. 228–229). Though they brought skill and enthusiasm to the project, the members of the NPL team had to struggle against technical and financial constraints. The Mark I was originally designed to have three packet switching nodes, but funding constraints reduced the number to one. With only one node, the NPL team would not get the opportunity to tackle certain issues—such as congestion and routing—that a multinode network would raise. Davies hoped that simulation studies could compensate for the lack of direct experience with a full-scale network (ibid., p. 229). The team had chosen for the node a computer, made by the English manufacturer Plessey, that had been designed specifically for data communications. After the NPL group had spent a year designing the Mark I around this machine, it was abruptly withdrawn from the market, and the team had to switch to a new computer. The chosen computer was the Honeywell 516.

A Honeywell 516 was installed at the NPL in 1969, and over the next two years user services were added to the network. The Mark I had about sixty lines that provided access to a DEC PDP-8 computer and two mainframes. Through the network, NPL researchers could have remote access to computers for writing and running programs, for querying a database, for sharing files, for special services such as a "desk calculator," and for "communication between people" (Davies 1966a, pp. 1–2). The system also included a file server and a "Scrapbook" application that provided document editing and communication tools (Campbell-Kelly 1988, p. 236).

Most of the effort centered on the design of the network interface. This design was shaped by Davies's assumption that businesspeople would turn out to be the main users of networks. Davies (1966b, p. 3) explained: "The emphasis on real-time business systems in this report is due to the belief that they will generate more real-time digital communication traffic than, say, scientific calculations or computer-aided design." Whereas academic researchers might need to transfer large amounts of data from one computer to another, businesspeople would be using terminals to access an interactive computer. The NPL designers therefore focused mainly on providing an easy-to-use terminal interface to the network.

One unusual characteristic of the Mark I that derived from the emphasis on user friendliness was that all terminals, printers, and other peripheral devices were connected directly to the network. The network was actually interposed between a computer and its own peripherals, so that the network became, in a sense, internal to the computer. Davies (1966b, p. 11) commented:

The overall description of the system shows a major organisational change. Present day multi-access computers each have equipment which assembles messages from keyboards and distributes them to printers. What we are proposing is that this function should be carried out by the network, not the attached computers.

Using the network as a common communication channel for all components would make it possible for any pair of machines to interact. Normally, a terminal user who wanted to print a file would have to log in to a host computer and send a command to a printer attached to that computer. With the Mark I, however, the user could send a command directly from their terminal to the printer, without ever having to go through the printer's host computer. Remote resources would be as easy to use as local ones, since the access procedures were identical. This was a radical concept in user interface design—a concept that would not become a commonplace feature of networked systems for another twenty years.

There was a price to pay for this vision, however. Since all terminals were connected through the network, a failure in the network would mean that terminals would be cut off even from their local host computer.[26] The variety of peripherals attached to the network also made the interface computer more complicated and expensive to build, which delayed the completion of the project.[27] And, in trying to make the terminal interface user friendly, the designers of the Mark I sacrificed flexibility and adaptability. For example, they implemented parts of the user interface in hardware (figure 1.4). A user wishing to set up a connection would punch a button marked TRANSMIT on the front of the terminal, after which a light labeled SEND would light up to indicate that the network was ready to accept data; there were other lights and buttons for different operations. This interface was easy for novices to learn, but it was harder to automate or modify than a procedure implemented in software would have been (anonymous 1967). In the fast-changing world of computing, a system that was not adaptable was in danger of becoming obsolete.

Figure 1.4
A Mark I terminal. The text on the screen reads "NPL Data Communications
Network." Source: National Physical Laboratory, Teddington. Reproduced by
permission of controller of HMSO.

The Impact of Davies's Work

The Mark I came to be used regularly by researchers at the National
Physical Laboratory, and in 1973 Donald Davies's team introduced an
upgraded version of the system called "Mark II." The Mark II used
most of the same hardware as the Mark I, but software improvements
made it two to three times faster. The Mark II remained in service at
the NPL until 1986—quite an impressive term of service for an experimental system (Campbell-Kelly 1988, pp. 237–239). Drawing on their
experience with the network, members of the NPL team went on to
participate in several larger network projects in the United Kingdom
and in Europe.

But despite Davies's technical innovations and the local success of
the system, the Mark I did not have the kind of influence that the
ARPANET would have. Davies was never able to build the national
network he had proposed, and the specific techniques used in the
Mark I were not transferred outside the NPL. Though Davies had had

a head start on the builders of the ARPANET, it was their work that would come to dominate the field of computer networking.

The politics of the day and the culture of some British institutions hampered Davies's ability to implement his ideas and fulfill his aim of keeping the United Kingdom ahead of the United States in computer networking. In the late 1950s the NPL had been oriented toward pure research, but under the Wilson government there was a marked increase in government oversight and intervention. In the recollection of one NPL scientist (Pyatt 1983, pp. 145–146):

Schemes for improving the service given to the nation were constantly being hawked from above. . . . Open-ended research was severely cut back and in its place all research projects had to have a 'customer,' who had to be persuaded of the viability and value of each project and agree to make available the funds to carry it out. . . . Meetings [with customers] required regular preparation of cases by Laboratory scientists in time which could ill be spared from practical work.

For Davies and the Mark I team, the emphasis on promoting commercial spinoffs of the network diverted time from actual research and development.

Another source of difficulty for the NPL was Mintech's attempt to "rationalize" the computer industry by forcing manufacturers to reduce the number of different types of computers they offered, on the theory that having a few models with large production runs would create economies of scale. Bowing to this policy, the Plessey Corporation canceled its plan to produce the minicomputer that the NPL team had chosen for its network interface. This delayed the NPL project and forced the NPL designers to make up for the lost functionality of the Plessey computer by increasing the complexity of other parts of the system.

Another major obstacle for Davies was that he needed help from the General Post Office (which had a monopoly on national telecommunications services) to build a large-scale network, and the GPO showed little interest in new computer technology. Davies was not alone in his vexation with the GPO. Early in 1967 a small but influential group of people involved in the British time sharing industry formed the "Real Time Club." This club's main activity was the sharing of information at informal monthly meetings, but it also occasionally lobbied the government to provide more support for data communications (Malik 1989; Foy 1986; Campbell-Kelly 1988, p. 228). Members of the Real

Time Club complained about the GPO's reluctance to provide better data communications:

The entrepreneurs discovered they were all hampered in their time sharing activities by the same thing—what they felt was foot-dragging on the part of the GPO . . . when it came to lines and modems for time sharing services. (Foy 1986, p. 370)[28]

Club members decided that public action was called for, and on 3 July 1968 they held a public event called "Conversational Computing on the South Bank" at London's Royal Festival Hall (Campbell-Kelly 1988, p. 228). Commercial time sharing firms demonstrated their services, leading figures in computing gave talks, and hundreds of computer professionals attended. One of the club's leading members, Stanley Gill, a professor at Imperial College, gave a speech urging that Donald Davies's network design be adopted. The Americans, Gill noted, were already working on plans for the ARPANET. Well attended and widely reported in the press, Conversational Computing on the South Bank generated a public debate on the idea of building a national packet switching network.

Eventually, the activism of computer users forced General Post Office authorities to develop data communications services. The GPO initiated several studies of networking, and with continued pressure from the Real Time Club the government began to give more support to networking research (Campbell-Kelly 1988, pp. 242–243).[29] The NPL's Roger Scantlebury, who had worked on the Mark I, helped supervise the research contracts for the GPO. In 1973 these activities led the GPO to begin work on its Experimental Packet Switching Service (EPSS), which became operational in 1977. However, though Davies's work had helped convince some influential people that a national network could and should be built, the design of EPSS differed significantly from Davies's vision of packet switching (ibid.).[30] Even worse from the NPL's perspective, the Post Office's next-generation Packet Switching Service was based on American rather than British technology; it used a system, developed by the American firm Telenet, that was a spinoff of the ARPANET project.[31] The Wilson government had aimed to encourage the development and exploitation of British computing technology, but its failure to coordinate decision making with the researchers on the front lines of innovation had had—at least in the case of the NPL networking effort—the opposite effect.

Putting It All Together: Packet Switching and the ARPANET

Paul Baran and Donald Davies had both envisioned nationwide networks that would use the new technique of packet switching, but neither man had been able to fully realize this goal. Instead, the first large-scale packet switching network would be built by the Advanced Research Projects Agency.[32] The design of this network would draw on the work of both Baran and Davies, but the network's builders had their own vision of what packet switching could achieve.

ARPA was one of many new American science and technology ventures that had been prompted by the Cold War. Founded in 1958 in response to Sputnik, ARPA had as its stated mission keeping the United States ahead of its military rivals by pursuing research projects that promise significant advances in defense-related fields.[33] Throughout its existence ARPA has remained a small agency with no laboratories of its own. ARPA managers initiate and manage projects, but the actual research and development is done by academic and industry contractors. Recognized even by its critics for good management and rapid development of new technologies, ARPA has had some success in transferring its technologies to the armed services and the private sector (Pollack 1989, p. 8).

The director of ARPA reports to the Director of Defense Research and Engineering at the Office of the Secretary of Defense. ARPA has several project offices that fund research in different areas; project offices are created or disbanded in response to the changing priorities of the Department of Defense. Each office has a director and several program managers, all of whom are directly involved in choosing research projects. The first project offices directed research in behavioral sciences, materials sciences, and missile defense. In 1962, with the founding of its Information Processing Techniques Office (IPTO), ARPA became a major funder of computer science in the United States, often outspending universities significantly. Computer science, not yet an established discipline in 1962, developed rapidly once IPTO began funding it. IPTO has been the driving force behind several important areas of computing research in the United States, including graphics, artificial intelligence, time sharing operating systems, and networking (Norberg and O'Neill 1996).[34]

ARPA's funding of basic research was consistent with the philosophy of the administration of President Lyndon Johnson, who, in a September 1965 memo to his cabinet, advocated the use of agency funds to support basic research in universities. In that memo, Johnson, noting that about two-thirds of universities' research spending was funded by various federal agencies, said that this money should be used to establish "creative centers of excellence" throughout the nation (Johnson 1972, p. 335). He urged each government agency engaged in research to take "all practical measures . . . to strengthen the institutions where research now goes on, and to help additional institutions to become more effective centers for teaching and research" (ibid., p. 336). Johnson specifically did not want to limit research at these centers to mission-oriented projects. "Under this policy," he wrote (ibid., p. 335), "more support will be provided under terms which give the university and the investigator wider scope for inquiry, as contrasted with highly specific, narrowly defined projects."

A few months later, the Department of Defense responded to Johnson's call with a plan to create "centers of excellence" in defense-related research. "Each new university program," the DoD suggested, "should present a stimulating challenge to faculty and students and, at the same time, contribute to basic knowledge needed for solving problems in national defense." (Department of Defense 1972, p. 337) IPTO created several computing research centers, giving large grants to MIT, Carnegie Mellon, UCLA, and other universities. By 1970, ARPA had funded a variety of time sharing computers located at universities and other computing research sites across the United States. The purpose of its proposed network—the ARPANET—was to connect these scattered computing sites.

The ARPANET project was managed by Lawrence Roberts, a computer scientist who had conducted networking experiments at MIT's Lincoln Laboratory before joining ARPA in 1966. Roberts had a mandate to build a large, multi-computer network, but he did not initially have a firm idea of how to do this. He considered having pairs of computers establish a connection using ordinary telephone calls whenever they needed to exchange data—a method he had employed in earlier experiments. But the high cost of long-distance telephone connections made this option seem prohibitively expensive. Roberts also worried that ordinary phone service would be unacceptably prone to transmission errors and line failures. Although he was aware of the concept of packet switching, Roberts was not sure how to implement it in a large network.

In October of 1967, with these issues still unresolved, Roberts attended a computing symposium in Gatlinburg, Tennessee, where he was slated to present ARPA's tentative networking plans. Roger Scantlebury of Britain's National Physical Laboratory also presented a paper at the symposium, where Roberts heard for the first time about Davies's ideas on packet switching and the ongoing work on the Mark I. After this session, a number of conference attendees gathered to discuss network design informally, and Scantlebury and his colleagues advocated packet switching as a solution to Roberts's concerns about line efficiency. The NPL group influenced a number of American computer scientists in favor of the new technique, and they adopted Davies's term "packet switching" to refer to this type of network. Roberts also adopted some specific aspects of the NPL design. For instance, Roberts had planned to use relatively low-speed telephone lines to connect the network nodes. He later recalled that, after the NPL representatives had "spent all night with [him] arguing about the thing back and forth," he had "concluded from those arguments that wider bandwidths would be useful" (Roberts 1989). Roberts decided to increase the bandwidth of the links in his proposed network from 9.6 to 56 kilobits per second. The ARPANET would also use a packet format similar to the NPL Mark I.[35]

After the ARPANET project was underway, the acoustics and computing firm of Bolt, Beranek and Newman, which had the main contract to build the network nodes, continued to interact with the NPL group. According to BBN's Robert Kahn (1990),

Donald Davies was a very creative guy; he thought a lot about interesting ideas of how networks should be built. He clearly had the concept in his head of what packet networks ought to look like, and he had done it independently in England. I believe Larry Roberts will probably tell you that Donald had a big influence on him.

The NPL's Derek Barber visited the BBN team in 1969; he reported that they "were interested in the possibility of connecting our type of local area [network] directly into" the ARPANET and that they saw the NPL work as "complementary" to the ARPANET project (Barber 1969, p. 15).[36]

Paul Baran, too, became directly involved in the early stages of planning the ARPANET. Roger Scantlebury had referred Lawrence Roberts to Baran's earlier work. Soon after returning to Washington from Gatlinburg, Roberts had read Baran's *On Distributed Communications*. Later he would describe this as a kind of revelation: "Suddenly I learned how to route packets." (Norberg and O'Neill 1996, p. 166)

Some of the ARPANET contractors, including Howard Frank and Leonard Kleinrock, were also aware of Baran's work and had used it in their research.[37] In 1967, Roberts recruited Baran to advise the ARPANET planning group on distributed communications and packet switching.

Through these various encounters, Roberts and others members of the ARPANET group were exposed to the ideas and techniques of Baran and Davies, and they became convinced that packet switching and distributed networking would be both feasible and desirable for the ARPANET. Packet switching promised to make more efficient use of the network's long-distance communications links and to enhance the system's ability to recover from equipment failures, which an experimental network would surely encounter. At the same time, however, packet switching was an unproven technique that would be difficult to implement successfully. The decision to employ packet switching on such a large scale reflected ARPA's commitment to high-risk research: if it worked, the payoff would be not only greater efficiency and ruggedness in the ARPANET itself, but also a significant advance in computer scientists' understanding of network properties and techniques. The ARPA managers could afford (indeed, had a mandate) to think extravagantly—to aim for the highest payoff rather than the safest investment.

The Social Construction of Packet Switching

The projects sponsored by Rand, the NPL, and ARPA had much in common in their approach to packet switching, but some crucial differences in ARPA's approach helped the ARPANET play a more enduring and influential role than the other projects. Donald Davies, Paul Baran, and Lawrence Roberts each made technical choices based on specific local concerns, and the extent to which their systems were influential depended in part on whether others shared those concerns. For instance, Baran's system had many elements that were specifically adapted to the Cold War threat, including very high levels of redundancy, location of nodes away from population centers, and integration of cryptographic capabilities and priority/precedence features into the system's design. None of these features were adopted by Davies or Roberts, neither of whom was concerned with survivability.[38] On the other hand, aspects of Baran's system that would be useful in a variety of situations—such as high-speed transmission, adaptive routing, and efficient packet switching—were adopted for use in later systems.[39]

One thing that Baran, Davies, and Roberts had in common was the insight that the capabilities of a new generation of small but fast computers could be harnessed to transcend the limitations of previous communications systems. Telephone systems of the late 1960s did not use computerized switches, and message switching systems used large, expensive computers that handled messages slowly. When presented with the idea that a network could employ dozens of computers as its switches, people in the communications industry tended to doubt that computers fast and cheap enough to make this idea feasible would be available (Baran 1990, pp. 19–21; Roberts 1978, p. 1307; Roberts 1988, p. 150; Campbell-Kelly 1988, p. 8). Indeed, the first of these small but powerful "minicomputers" did not appear until 1965, when the Digital Equipment Corporation introduced its PDP-8. The fact that packet switching relied on an innovative computer product helps to explain why that technique was consistently explored by computer scientists but not by communications experts, even though it drew on aspects of both fields.

In the 1960s, computing technologies became policy instruments both in the United States and in the United Kingdom. In the United Kingdom, intervention in the computer industry was seen as a symbol of the Labour Party's commitment to modernization and as an engine of economic growth, and the government made efforts to fund research and coordinate industrial production. In the United States, technological prowess was seen as a weapon in the Cold War, and defense-related research was generously funded through organizations such as the Rand Corporation and ARPA. In both countries, individuals and organizations interested in pursuing computer networking often found it necessary to join government-sponsored projects or to present their work as responsive to contemporary political agendas.

Although computer networking had a political role in both countries, there were striking differences in the levels of government funding, in policy makers' interpretation of networks as a military or a civilian technology, and in government's inclination to intervene in private enterprise. These differences are evident in the contrasting outcomes of the attempts by the NPL and ARPA to build large-scale networks. The United States poured much more money into basic computing research than did the United Kingdom, and most of that money was channeled through the Department of Defense. Not only did Roberts have a generous budget for his project; he also was able to call on computer experts from around the country to help build the network. Davies, at the NPL, had a much smaller budget. Faced with a perceived economic crisis and convinced of the need to compete with the United States and other exporters of high technology, the British government tried to rationalize the computing industry and to encourage commercial spinoffs of government research. Eventually much of the research at the NPL and at similar places was directly focused on short-term commercial applications, and the Labour government's industrial policy limited Davies's choice of computers. The US government was less inclined to try to manage the domestic computer industry. Overall, Roberts had much more support and much less interference from his government than Davies had from his.

Davies had been one of the earliest and most articulate advocates of packet switching. He had formulated a detailed plan for a national network at a time when the ARPANET was still just an idea. Yet by the middle of 1968 Davies was already lamenting that his project had been eclipsed by the American effort: "As a force in this discussion NPL is too remote and our own demonstration as planned now is small-scale and likely to be delayed by the reductions in staff and administrative difficulties in purchasing computers." (Davies 1968b, p. 7) Despite their technological vision, neither Baran nor Davies could find the backing to build a national packet switching network. Roberts, in contrast, was able to make the ARPANET an internationally recognized symbol of the feasibility of packet switching only a few years after he learned of the technique.

The fact that packet switching had to be integrated into local practices and concerns led to very different outcomes in the three network projects. Some visions of packet switching were easier to implement, some turned out to be a better match for evolving computer technology, and some were more attractive to organizations in a position to sponsor network projects. Making packet switching work was not just a matter of having the right technical idea; it also required the right environment. Only after the ARPANET presented a highly visible example of a successful packet switching system did it come to be seen as a self-evidently superior technique. The success of the ARPANET may have depended on packet switching, but it could equally well be argued that the success of packet switching depended on the ARPANET.

2
Building the ARPANET: Challenges and Strategies

The ARPANET was born from an inspiration and a need. The inspiration can be traced back to Joseph C. R. Licklider, the first director of ARPA's Information Processing Techniques Office (IPTO). Licklider's influential 1960 paper "Man-Computer Symbiosis" became a manifesto for reorienting computer science and technology to serve the needs and aspirations of the human user, rather than forcing the user to adapt to the machine. Licklider (1960, pp. 4–5) wrote:

> The hope is that, in not too many years, human brains and computing machines will be coupled together very tightly, and that the resulting partnership will think as no human brain has ever thought and process data in a way not approached by the information-handling machines we know today. . . . Those years should be intellectually the most creative and exciting in the history of mankind.

Licklider went on to identify specific changes in the practice of computing that were needed to bring about "symbiosis," including interactive computers, more intuitive methods for retrieving data, higher-level programming languages, better input and output devices, and data communications. As director of IPTO, Licklider funded the development of time sharing systems, which made interactive computing economically feasible for large numbers of users. To many computer scientists, networking seemed like the next step in interactive computing and a logical extension of time sharing. In an early ARPANET paper, Roberts and Wessler (1970, p. 543) reasoned: "Within a local community, time sharing systems already permit the sharing of software resources. An effective network would eliminate the size and distance limitations on such communities."

Among those who heeded Licklider's message was Robert Taylor, who had been a systems engineer in the aerospace industry and at

NASA before joining ARPA in 1965.[1] By the mid 1960s IPTO was funding computing research centers around the country to work on projects such as time sharing, artificial intelligence, and graphics. Taylor (1989) felt that each of IPTO's scattered research centers "had its own sense of community and was digitally isolated from the other one." When he became director of IPTO, in 1966, Taylor began to speculate on ways to "build metacommunities out of these by connecting them" (ibid.). Early that year, he and ARPA director Charles Herzfeld discussed a plan to link IPTO's computing sites with an experimental network. In 1967 Herzfeld agreed to allocate $500,000 for preliminary work on the idea, which was dubbed the ARPA Network or ARPANET. Figures 2.1 and 2.2 illustrate how the ARPANET reproduced the geography of ARPA's research network, spanning the United States to link ARPA's computing sites.

Besides serving Taylor's vision of linking the research community, the network would address a pressing need within ARPA. ARPA was the major funding source for most of its computing contractors, and buying computers for them represented a large expense for the agency. To make matters worse, a single contractor might need access to several types of machines. Computer hardware and operating systems tended to be optimized for particular uses, such as interactive time sharing or high-powered "number crunching." Computers also had a variety of specialized input/output devices, such as graphics terminals. Contractors who wanted to combine different modes of computing had to either travel to another site or acquire multiple machines. As a result, IPTO was continually besieged by requests from its contractors for more computers. Taylor believed that if ARPA's scattered computers could be linked together, hardware, software, and data could be efficiently pooled among contractors rather than wastefully duplicated.

In late 1966 Taylor recruited Lawrence Roberts, a program manager at MIT's Lincoln Laboratory, to oversee development of the ARPANET. Roberts had been pursuing networking experiments at the Lincoln Lab, and Taylor considered him the best qualified candidate to manage the ARPANET project, but Roberts was initially reluctant to leave his research position. The circumstances of his joining IPTO provide an example of ARPA's leverage over the computer science research community. When Roberts turned down an initial invitation to come to ARPA, Taylor asked ARPA's director, Charles Herzfeld, to call the head of the Lincoln Lab and remind him that half of his lab's funding came from ARPA, and that it would be in the lab's best

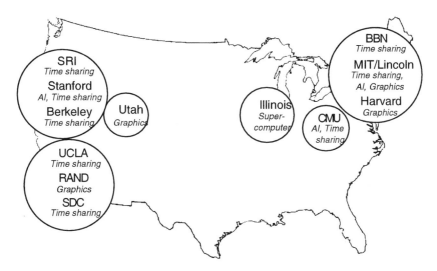

Figure 2.1
The main IPTO research centers at the time of the ARPANET's creation.

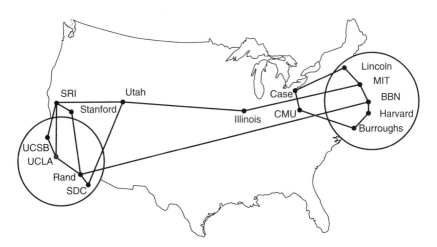

Figure 2.2
A map of the fifteen-node ARPANET in 1971, redrawn from Bolt, Beranek
and Newman's original. (SDC: Systems Development Corporation. CMU:
Carnegie Mellon University.)

interests to send Roberts to Washington.[2] Roberts joined IPTO as assistant director and became director when Taylor left the agency in March of 1969.

Roberts envisioned the ARPANET as a way to bring researchers together. He stressed early on that "a network would foster the 'community' use of computers." "Cooperative programming," he continued, "would be stimulated, and in particular fields or disciplines it will be possible to achieve a 'critical mass' of talent by allowing geographically separated people to work effectively in interaction with a system." (Roberts 1967b, p. 2) Roberts also saw the network as a chance to advance the state of the art in computer science. As he knew from his own experience in the field, networking techniques were still in a rudimentary stage, and many theoretical and practical questions remained unexplored.

For ARPA's managers, then, the network project represented a chance to pursue advanced research in a new branch of computer science, potential financial savings for the agency, and the fulfillment of a vision of interactive computing. These goals set the general outline of the proposed network. As we saw in chapter 1, Roberts decided that the network should be a distributed packet switching system, so as to reduce transmission costs, increase reliability, and potentially further the military objective of developing sophisticated and survivable communications systems. The network would extend across the United States, matching the distribution of ARPA sites. It would link time sharing computers to support both remote terminal access to distant computers and high-volume data transfers between computers. Since the network represented an experiment in data communications techniques, the scope of the project would include not only building the system but also testing and analyzing its performance. Finally, in order to maximize the resources available and to save ARPA money on computer facilities, Roberts required all IPTO sites to participate by connecting their computers to the network—whether they wished to or not.

Roberts bore most of the responsibility for seeing the ARPANET project through to a successful conclusion, and his management skills proved invaluable. Building a long-distance packet switching network to connect diverse computers would be a formidable task, even for an agency with ARPA's resources and its mandate for advanced research. Beyond its sheer size, the ARPANET was one of the most complex computing projects of its time, pushing forward the state of the art in data communications. To keep the project on track, Roberts deployed a unique set of technical and managerial strategies. Both the ARPA-NET itself and ARPA's approach to building it would have a lasting influence on the emerging field of computer networking.

Initial Challenges

As Taylor, Roberts, and other members of the ARPA computer science community began working out the design of the ARPANET, it became clear that building a network according to their specifications would present enormous technical challenges. Packet switching was a risky choice for the ARPANET; using this novel technique would increase the uncertainty and complexity of the system design and hence the project's chances of failure. In 1967 the world's first packet switching computer network was still in the planning stages (at Britain's National Physical Laboratory), and many experts were openly skeptical that such a system could work. The fact that the eventual success of ARPANET was widely interpreted as a proof of the feasibility of packet switching indicates that the technique had not previously achieved wide acceptance. Roberts (1988, p. 150) found that telephone engineers questioned his credibility for even suggesting such a radical departure from established practice: "Communications professionals reacted with considerable anger and hostility, usually saying I did not know what I was talking about." Communications experts were familiar with the difficulty of routing messages individually through a network, and it was clear to them that breaking messages into packets would add to the complexity of the system. From their perspective, the activities required of packet switching nodes seemed too difficult to be performed quickly, reliably, and automatically. The communications experts at the Defense Communications Agency were no more sympathetic, according to Roberts (1989). Even within the field of computer science, critics pointed out difficulties. Packets sent through the ARPANET would have to be reordered and reassembled into complete messages at their destinations. Some experts predicted that this would require excessive amounts of computer memory. Others argued that a routing system that changed rapidly in response to traffic conditions might send packets looping endlessly through the network (Rinde 1976, p. 271; Roberts 1978, pp. 1307–1308). Some of these problems did, in fact, occur in the ARPANET, and they took considerable effort to fix.

Another unusual and potentially troublesome characteristic of the ARPANET was the great variety of computers it would connect. Besides machines commercially available from IBM, DEC, GE, SDS, and UNIVAC, the proposed ARPANET sites had various one-of-a-kind machines, such as ARPA's experimental ILLIAC supercomputer (Dickson 1968, p. 132). These various types of computers were incompatible with one another, which meant that users who wanted access to programs or data at other sites often had to reprogram the software or reformat the data. In 1969, ARPA director Eberhardt Rechtin told Congress: "When one user wants to take advantage of another's developments, he presently has little recourse except to buy an appropriate machine or to convert all of the original software to his own machines." (US Congress 1969, p. 809) Incompatibility wasted time and programming resources, and it remained an obstacle to collaborative work. For Roberts, one aim of the ARPANET project was to overcome these obstacles. Roberts viewed the diversity of computers not as an unfortunate necessity but as a strength of the system, since a network that connected heterogeneous systems could offer users a wider range of resources. But getting this assortment of machines to communicate would require an enormous effort in hardware and software redesign. "Almost every conceivable item of computer hardware and software will be in the network," Roberts pointed out, adding "This is the greatest challenge of the system, as well as its greatest ultimate value." (quoted in Dickson 1968, p. 131)

Roberts's view was based on his experience as one of the first people to attempt to establish a connection between different types of computers. After receiving his Ph.D. in Electrical Engineering from MIT in 1959, Roberts began working at the Lincoln Laboratory, where he became interested in the possibility of networking computers for time sharing during discussions with J. C. R. Licklider, Donald Davies, and others in 1964 and 1965 (Roberts 1988, pp. 143–144). Roberts found a kindred spirit in Thomas Marill, who had studied under Licklider and had founded a time sharing company in Cambridge called the Computer Corporation of America. In 1966, with funding from IPTO, Roberts and Marill undertook to build a rudimentary network linking two experimental computers: the TX-2 at the Lincoln Lab and the Q-32 at the System Development Corporation in Santa Monica. A line leased from Western Union provided the communications link, and Marill and Roberts wrote their own software to manage the connection. They published their results in the fall of 1966, just before Roberts left Lincoln Lab for ARPA.

In describing their experiment, Marill and Roberts articulated some important concepts. In their view, the "elementary approach" to connecting two computers was for each computer to treat the other as a terminal. Such a connection required little modification of the computers, but it had severe limitations. The connection was slow, since terminals operate at much lower data rates than computers, and there was no general-purpose way to access a remote system, since each separate application program had to manage its own connections rather than having the operating system handle the connections for all applications. Marill and Roberts thought that forgoing the elementary approach and taking on the harder task of modifying the computers' operating systems would make it possible to create a higher-speed computer-to-computer interface instead of relying on the ordinary terminal-to-computer interface. They proposed that each host computer implement a general-purpose set of rules for handling a network connection, which they called the "message protocol" (Marill and Roberts 1966, p. 428). Roberts applied what he had learned from this experiment to the design of the ARPANET. He decided that all the host computers should follow a standard protocol for network interactions. Having a standard protocol would help overcome the incompatibilities between different types of computers. However, this approach also created a huge task for the people maintaining the hosts, who would have to add this new networking capability to the operating systems of their various computers.

Creating a heterogeneous, packet switching, continent-spanning computer-to-computer network would be a significant technical achievement for ARPA; the challenge would lie in keeping these same features from leading the project into chaos. The technical and managerial difficulties of the ARPANET project became apparent when Taylor and Roberts presented the network concept at IPTO's annual meeting of Principal Investigators (scientists heading research projects) at the University of Michigan in April of 1967. Roberts had already discussed the idea informally with several of the PIs, but at the meeting he announced that the project would definitely go forward. The PIs, who would have to design, implement, and use the proposed network, did not greet the network idea with the enthusiasm it would receive in later years. Most PIs at the meeting reacted with indifference or even hostility to the idea of connecting their computer centers to the network. Some of them suspected—correctly—that ARPA saw the network as an alternative to buying them more computers. Roberts (1989) recalled:

Although they knew in the back of their mind that it was a good idea and were supportive on a philosophical front, from a practical point of view, they—Minsky, and McCarthy,[3] and everybody with their own machine— wanted [to continue having] their own machine. It was only a couple years after they had gotten on [the ARPANET] that they started raving about how they could now share research, and jointly publish papers, and do other things that they could never do before.

Many PIs did not want to lose control of their local computers to people at other sites, and they saw the network as an intrusion.[4] Since "their" machines were actually paid for by ARPA, the PIs had little choice in the matter; however, they were not eager to join in the network. Even those who agreed on the general advantages of developing computer networks had practical objections to implementing the ambitious system envisioned by Roberts and Taylor. Some of these PIs were unwilling to undertake the massive effort that seemed to be required; others were convinced that the project would fail altogether.

Besides reminding us that even those at the forefront of computer science in 1967 could not foresee the astounding popularity of the ARPANET and its successors, the negative reactions of the Principal Investigators illustrate the two major challenges that ARPA faced. First, it was clear that the complexity of the network's design would require imaginative technical solutions. Second, ARPA would need to find ways to gain the cooperation of prospective network members. The PIs were initially more concerned with continuing their own local projects than with collaborating on a network. In order for the project to succeed, Lawrence Roberts would need to create some sense of common purpose.

System-Building Strategies

Of the many problem-solving strategies that Roberts and his team of contractors would employ in building the ARPANET, two were especially significant. One was an approach that came to be known as *layering*, which involved dividing complex networking tasks into modular building blocks. The second was an informal and decentralized management style. Layering and a decentralized, collegial approach to management came to be seen by members and observers of the project as essential characteristics of the ARPANET, and were later held up as models for successful project development; this gave these techniques an influence beyond their role as management tools for the ARPANET project.[5]

Layering

A layered system is organized as a set of discrete functions that interact according to specified rules. The functions are called "layers" because they are arranged in a conceptual hierarchy that proceeds from the most concrete and physical functions (such as handling electrical signals) to the most abstract functions (e.g., interpreting human-language commands from users). Each higher-level function builds on the capabilities provided by the layers below. The idea of layering seems to have occurred independently to many people working on networks as they drew on concepts of modularity and functional division of systems that were current in computer science.[6]

In the ideal layered system, the opportunities for interaction among layers are limited and follow set rules. This reduces the complexity of the system, making it easier to design, test, and debug. The designer of a particular layer needs to know how that layer is expected to interact with other layers but does not need to know anything about the internal workings of those layers. Since the layers are independent, they can be created and modified separately as long as all those working on the system agree to use the same interfaces between layers. Thus, layering has both technical and social implications: it makes the technical complexity of the system more manageable, and it allows the system to be designed and built in a decentralized way.

The ARPANET's builders did not start out with a specific plan for how functions would be divided up among layers or how the interfaces and protocols would work. Rather, a layered model evolved as the ARPANET developed. The first step toward a layered approach was taken at the 1967 meeting of Principal Investigators. One of the contractors' main concerns on first hearing about the project was that creating the necessary packet switching software for their computers would require too much effort on their part. IPTO research sites used a wide variety of time sharing operating systems; if the host computers had to perform packet switching, someone would have to program each different type of computer to perform the various packet switching tasks and then reprogram each computer whenever the software needed modification. Moreover, the packet switching software would have to be designed to accommodate the limitations and idiosyncrasies of each model of computer. In view of these difficulties, even PIs who were sympathetic to the project's goals had reason to be skeptical about its technical feasibility.

One of the Principal Investigators, Wesley Clark of Washington University in St. Louis, saw an easier alternative. Clark was familiar with the capabilities of minicomputers, and after the meeting he suggested to Roberts that each of the host computers be attached to a special minicomputer that would act as the host's interface to the network. In Clark's plan, the minicomputers, rather than the hosts, would form the nodes of the network and handle the packet switching operations. This network of minicomputers was designated the *subnet*. Since minicomputers were becoming relatively inexpensive by the late 1960s, it seemed economically feasible to dedicate several of them to running the network. Taylor endorsed the subnet scheme, and Roberts incorporated it into the ARPANET design's, calling the minicomputers "interface message processors" (IMPs).[7] Figure 2.3 illustrates the subnet idea.

The subnet design created a division of labor between the switching nodes (IMPs), whose task was to move packets efficiently and reliably

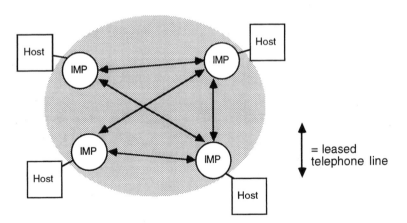

Figure 2.3
Network model with communications subnet.

Table 2.1
The two-layer model of the ARPANET.

Layer name	Functions
Host	Handles user interface; initiates and maintains connections between pairs of hosts
Communications	Moves data through subnet using packet switching; ensures reliable transmission on host-IMP and IMP-IMP connections

from one part of the network to another, and the hosts, which were responsible for the content of those packets. Packet switching programs could now be written for a single type of IMP computer rather than many different types of hosts. Host administrators could treat the entire subnet as a "black box" that provided a service without requiring them to know how it worked, and could focus their energies on providing host resources. The ARPANET team began to see the system as being divided conceptually into two layers: a communications layer, consisting of packet switching IMPs connected by leased telephone lines, and a host layer, which would coordinate interactions between host processes and provide user services (Heart et al. 1970, p. 551). This model is summarized in table 2.1.

In depicting the network as a "stack" of layers or protocols, the two-layer model (table 2.1) suggests two kinds of relations between system functions. First, the functions become increasingly abstract as one moves from the bottom to the top of the stack—from moving electrons over wires to interpreting commands typed by terminal users. Second, the order of the layers represents a temporal sequence from top to bottom: first the user types a command that invokes the host program, then the host protocol sends packets to the communications subnet.

The "protocol stack" model would quickly come to dominate the way people thought about organizing networks precisely because it offered a blueprint for reducing the complexity of network components while increasing the predictability of the system as a whole.[8] Before the ARPANET was finished the model would be expanded to three layers, and in later years still more layers would be added to keep pace with new capabilities and new ideas about how to organize networks.

Informal Management

Whereas the layering approach stressed separating the system's elements, ARPA's management style was aimed at fostering the cooperation required to integrate those elements into a coherent whole. ARPA's unmatched financial resources drew many computer scientists into its projects, but ARPA managers did not conduct relations with their researchers on a purely financial, contractual basis. The organizational culture surrounding the ARPANET was notably decentralized, collegial, and informal. In coordinating its contractors, ARPA relied largely on collaborative arrangements rather than contractual obligations, and technical decisions were usually made by consensus. The network itself provided a new way to coordinate dispersed activities and came to function as a meeting place for the computer science community. Though conflicts sometimes arose among the contractors, the ARPANET culture enhanced ARPA's ability to enlist the support of the research community and to respond to the technical challenges that the project posed.

The collegial management style of Taylor and Roberts was typical of IPTO in the 1960s and the 1970s. IPTO recruited most of its directors and project managers from the ranks of active researchers at university and industrial research centers. IPTO managers kept in touch with their colleagues by touring contract sites to evaluate the progress of programs, learn about new ideas, and recruit promising researchers. Not career managers, they generally stayed at ARPA only a few years before returning to academia or private business (in part because ARPA salaries were modest). Though ARPA as an organization had financial power over its contractors, most of the individuals who actually managed IPTO projects were drawn from those contractors. Howard Frank of the Network Analysis Corporation, an ARPANET contractor, observed: "It's easy to say 'the government,' or ARPA, or something like that, but they are individuals that you deal with." (Frank 1990, p. 300)

ARPA tended to award contracts through an informal process, funding individuals or organizations who were already known to IPTO managers for their expertise in a particular area. The ARPA approach exhibited the weaknesses and the advantages of an "old boy" network. Many talented computer scientists found themselves left out of the field's biggest funding opportunity, but those who were included enjoyed an extremely supportive environment.[9] Wesley Clark (1990) commented: "In the ARPA system, once you were in, you were a member of a club, . . . with . . . a pretty good sense of community with other people who were receiving support from that office." Robert Taylor made a special point of providing ongoing funding for graduate students at contract sites, and he arranged special meetings and working groups for them (Taylor 1989, p. 19). Graduates of the IPTO-funded programs at MIT, Stanford, Carnegie Mellon, and elsewhere became a major source of computer science faculty at American universities, thereby extending ARPA's social network into the next generation of researchers (Norberg and O'Neill 1996, pp. 290–291).[10]

In view of the expense of computing machinery in the 1960s and ARPA's large role in funding computer science, IPTO managers had real power over their contractors, and they were willing to use this power when they felt it necessary. As has already been noted, Robert Taylor exerted pressure on Lawrence Roberts to leave his position at the Lincoln Lab and join ARPA. Once in charge of the project, Roberts did not hesitate to make reluctant contractors share in the ARPANET effort:

> The universities were being funded by us, and we said, "We are going to build a network and you are going to participate in it. And you are going to connect it to your machines. By virtue of that we are going to reduce our computing demands on the office. So that you understand, we are not going to buy you new computers until you have used up all of the resources of the network." So over time we started forcing them to be involved. (Roberts 1989)

But IPTO managers preferred to take the informal approach whenever possible. Having been researchers themselves, they subscribed to the view that the best way to get results in basic research was to find talented people and give them room to work as they saw fit. They also tended to believe that differences of opinion could be debated rationally by the parties involved and decided on their technical merits, and that they, as IPTO managers, would need to intervene with an executive decision only if the contractors could not resolve differences among themselves. Not surprisingly, IPTO contractors praised this management style as an enlightened and productive way to conduct research. The report of an outside consultant commissioned by ARPA to report on the project's status in 1972 agreed that the project's informal style had contributed to its success, noting that the process of building the ARPANET had "been handled in a rather informal fashion with a great deal of autonomy and an indefinite division of responsibilities among the organizations that address the various elements of this function." The report continued: "Personal contacts, telephone conversations, and understandings are relied upon for day to day operation. This environment is a natural outcome of the progressive R&D atmosphere that was necessary for the development and implementation of the network concept." (RCA Service Company 1972, p. 34)[11]

In view of the nature of the project, it made sense for Roberts to encourage ARPA's network contractors to work together as peers. Different tasks required different combinations of skills, and no one contractor had the overall expertise or authority to direct the others as subordinates. Roberts's informal coordination methods provided a context in which the network builders could, for the most part, share skills and insights on an equal and cordial basis.

Getting Started

Roberts began the ARPANET project informally. Rather than soliciting bids for contracts right away, he brought together a small group of Principal Investigators who had expressed interest in the network concept and began meeting with them to discuss design problems and to work out possible solutions. He asked Elmer Shapiro of the Stanford Research Institute to lead these meetings, and he recruited Paul Baran of the Rand Corporation (the man who had done the earliest work on distributed communications and packet switching) to advise the group.

The various members of this group incorporated their own values into the ARPANET's design. For instance, while time sharing enthusiasts insisted on very fast response, so that users would not be frustrated by long delays, more analytically oriented researchers such as UCLA's Leonard Kleinrock insisted on incorporating measurement software into the switches so that they would be able to study the network's performance (Kleinrock 1990). The institutional homes of the members of this self-selected group would also become the first four nodes of the network: the University of California at Santa Barbara, the Stanford Research Institute, the University of Utah, and the University of California at Los Angeles. In June of 1968, Roberts submitted the plan his group had worked out to ARPA director Herzfeld. In July, Roberts received an initial development budget of $2.2 million and approval to seek contractors to develop the network.[12]

The basic infrastructure of the ARPANET would consist of time sharing hosts, packet switching interface message processors, and leased 56-kilobits-per-second telephone lines to connect the IMPs. The hosts were already in place, and the lines would be provided by AT&T,

so the main development task was to build the IMPs. Unlike most IPTO projects, which were initiated by contractors who had already shown interest and expertise in a given area, building a packet switching computer was a new venture initiated by ARPA, and there was no obvious candidate for the job. Therefore, Roberts departed from ARPA's usual practice and solicited competitive bids for the IMP contract from a number of computer and engineering firms. In early 1969, after considering bids from a dozen companies of all sizes, Roberts awarded the contract to the Bolt, Beranek and Newman Corporation of Cambridge, Massachusetts, a relatively small company specializing in acoustics and computing systems.

Though not a giant in the computer business, Bolt, Beranek and Newman had several advantages behind its bid. The company had previous ties with IPTO: J. C. R. Licklider had worked at BBN before becoming the first director of IPTO, BBN had contributed to IPTO's earlier time sharing efforts, and BBN researcher Robert Kahn had discussed networking with Roberts during the early stages in the planning of the ARPANET. BBN was also known for its strength in research. In 1990, Kahn recalled BBN as having been "a kind of hybrid version of Harvard and MIT in the sense that most of the people there were either faculty or former faculty of either Harvard or MIT" and as "sort of the cognac of the research business, very distilled" (Kahn 1990, p. 11). Frank Heart, Severo Ornstein, David Walden, and William Crowther of BBN's IMP team had all worked at the Lincoln Lab, where they had acquired valuable experience with "real-time" computer systems, which process and respond to input as fast as they receive it. Programmers of real-time systems must learn to write software that is compact and efficient—skills that the BBN team would need to ensure that the IMP subnet would provide responsive communications for interactive computing. Finally, BBN had valuable ties to the Honeywell Corporation, whose H-516 minicomputer was a strong candidate for the IMP, being fast, economical, well tested in actual use, easily programmed, and equipped with good input/output capabilities (Heart et al. 1970, p. 557). BBN and Honeywell were located conveniently near each other in the Boston area, and the two companies had agreed to work together on customizing the H-516 for use in the network if BBN's bid were accepted.

Other contracts were awarded less formally. Since the network was supposed to be a research experiment as well as a communications tool, Roberts awarded a contract to Leonard Kleinrock of UCLA to create theoretical models of the network and to analyze its actual performance. Roberts knew Kleinrock from MIT, where they had been graduate students together, and he was aware that Kleinrock had adapted a mathematical tool called queuing theory to analyze network systems. UCLA was designated the Network Measurement Center, and Kleinrock and his students would take an active part in testing and refining the network. To allow the analysis to begin as soon as possible, UCLA was chosen as the site of the first IMP.

Once the initial four-node network was functioning smoothly, Roberts planned to extend the ARPANET to fifteen computer science sites funded by IPTO (figure 2.2), then to additional ARPA research centers doing work in other fields, and then perhaps to military sites (Roberts and Wessler 1970, p. 548). With fifteen or more sites, it would be no simple matter to design a network that combined the redundant connections needed for reliability with maximum data throughput and minimum cost. Thus, Roberts contracted the Network Analysis Corporation (headed by Howard Frank) to help in planning the network's topology—the layout of nodes and communications links. Frank had met Kleinrock when they were both lecturing at Berkeley, and Kleinrock had later introduced Frank to Roberts. Frank, whose experience included optimizing the layout of oil pipelines, had formed NAC to provide consulting services for businesses building complex systems. For Frank, taking part in ARPA's cutting-edge research project was a welcome change from more routine commercial work, and it was also a way to keep his employees motivated: "It was the more intellectually challenging of the work. And I could have really smart, capable people working on that . . . where, if you only put them into commercial applications, after a while they would just leave. They would get very burned out." (Frank 1990) Frank's team would use recently developed computer simulation techniques to evaluate the cost and performance of various network topologies. NAC's analytical tool was a "heuristic" computer program—one that provides an approximate solution to a problem whose exact solution would require too much computing time. They would give as input to the program an initial topology that satisfied the performance constraints specified by ARPA: a maximum delay of 0.2 second for message delivery (for responsiveness), a minimum of two links per IMP (for reliability), and easy expandability. The program would then systematically vary small portions of this topology, rejecting changes that raised costs or violated constraints and adopting changes that lowered costs without violating constraints. By running the program thousands of times using different starting topologies and constraints, NAC was able to provide a range of solutions that satisfied ARPA's performance criteria while minimizing costs (Frank, Frisch, and Chou 1970, p. 581).

Roberts gave the Stanford Research Institute a contract to create an online resource called the Network Information Center, which would maintain a directory of the network personnel at each site, create an online archive of documents relating to the network, and provide information on resources available through the network (Ornstein et al. 1972, p. 253). In addition to Elmer Shapiro and a number of other computer scientists who were interested in the network, SRI had on its staff Douglas Engelbart, a pioneer of human-computer interface design who, in 1965, had invented the mouse. Engelbart was developing a database, a text-preparation system, and a messaging system with a sophisticated, user-friendly interface. Roberts hoped that, with such tools available, SRI would provide an easy-to-use information center for the network community.

Roberts also reestablished his informal networking group, now named the Network Working Group (NWG), to develop software specifications for the host computers and to provide a forum for discussing early experiences and experiments with the network. The most active members of this group were computer science graduate students who had been asked by their advisers to represent their sites. At UCLA, which was particularly active in the NWG, Leonard Kleinrock was using ARPA money to support a number of Ph.D. students, including Stephen Crocker, Vinton Cerf, and Jon Postel, all of whom were to be important figures in the development of host software. Other active members in the early stages of the NWG included Jeff Rulifson, Bill English, and Bill Duvall at SRI, Gerard Deloche at UCLA, Steve Carr at the University of Utah, Ron Stoughton at UC Santa Barbara, John Haefner at Rand, Bob Kahn and Dave Walden at Bolt, Beranek and Newman, and Abhay Bhushan at MIT.[13] The membership of the group changed constantly, growing larger as more sites got involved in the network. The NWG would gradually develop a culture of its own and a style of approaching problems based on the needs and interests of its members.

Most of the contractors had previous social connections with ARPA personnel or contractors. Kleinrock, as has already been noted, brought Frank into the project. In addition, Roberts, Kleinrock, Heart, and Kahn had all earned degrees at MIT, and Roberts,

Table 2.2
The organization of the ARPANET project.

ARPA/IPTO *Project Management*				
Bolt, Beranek and Newman	University of California at Los Angeles	Network Analysis Corporation	Stanford Research Institute	Network Working Group
IMP hardware and software, network operations	*Analysis*	*Topology*	*Network Information Center*	*Host protocols*
Honeywell *IMP hardware*				

Kleinrock, and most of the members of the BBN team had worked and become acquainted at the Lincoln Lab. The informal atmosphere of the project was, no doubt, attributable in large part to the many social ties among the contractors.[14]

Table 2.2 illustrates the organization of the ARPANET project. The distribution of contracts followed the layered division of the network itself. BBN, UCLA, and NAC worked on the communications layer, while the NWG designed the host software and SRI provided documentation services for the host layer. Having the various layers perform independent functions made it easier for ARPA to distribute the development work among several groups.

Managing Technical Complexity

Building the IMP: Putting Layering into Practice
At the heart of the communications subnet was the interface message processor, which acted both as a packet switch and as an interface between the host and the network. Since it was to be an interface, the IMP had to make data from hosts conform to the packet format used in the subnet. The task was relatively simple. An IMP would receive data from hosts in the form of "messages" that contained up to 8096

bits. The IMP would break up these messages to fit into packets of about 1000 bits, then add to each packet a standard header containing the packet's source and destination addresses and some control information that could be used to check for transmission errors. When the packets reached their destination, the IMP there would strip off the packet headers and reassemble the packets into a complete message before handing the data over to the host.

As a packet switch, the IMP had to ensure that data was transmitted efficiently and reliably along each link between a pair of IMPs or between an IMP and a host. One mechanism for increasing reliability was to acknowledge receipt of the packets. Whenever an IMP or a host sent a packet across a link, it waited for the recipient to send back a standard message indicating that the data had been received intact. If this acknowledgment did not arrive within a given period of time, the sender would transmit the packet again. Before acknowledging receipt of a packet, the IMP used a mechanism called a "checksum" to verify that the data had not been corrupted during transmission.[15] Acknowledgments and checksums used the computing power of the IMP to give links a degree of reliability that many telephone engineers, on the basis of their experience with analog systems, had thought unattainable. The IMP was also responsible for controlling the flow of traffic over the network to prevent congestion. The BBN team initially tried to accomplish this by having the IMP restrict the number of packets any host could send into the network at one time; however, designing good flow-control mechanisms proved to be a difficult task that computer scientists were still wrestling with decades later.

Perhaps the most difficult packet switching task for the IMP was routing. In the ARPANET, routing was distributed: rather than having a central routing mechanism, each IMP decided independently where to send packets.[16] To find the shortest routes, the IMP kept a table with an entry for each host on the network, showing how long it would take a packet sent from the IMP to reach that host and which of the IMP's links led to that host by the fastest route. When a packet came in, the IMP would look up the destination host in the table and forward the packet via the specified link. The routing system was also notable for being adaptive, continually responding to changes in network configuration or traffic. Every 2/3 second, the IMP would make a new estimate of how long it would take to reach the various host destinations, and it would send these routing estimates to each of its neighbors. The IMP used the information sent in by its neighbors to update its own routing table, matching each host destination with the link that had reported the shortest travel time to that host. This innovative approach to routing served ARPA's goal of building a rugged, flexible system. Distributed routing made the system more robust by minimizing its dependence on any one component. Adaptive routing allowed IMPs to improve the speed and reliability of the network by avoiding congested routes and node or line failures. The price of relying on so many independent, constantly changing routing decisions, however, was a complex system prone to unexpected interactions. As a result, Bolt, Beranek and Newman's IMP team had to revise its routing algorithm several times as experience or simulation revealed weaknesses (Ornstein et al. 1972, p. 244; Stallings 1991). The ARPANET's approach to routing reflected its designers' commitment to exploring new techniques and building a high-performance network, even at the price of creating a system that was, at times, difficult to understand and control.

The interface and packet switching functions just described had been specified by the ARPA contract. But the design of the IMP was also shaped by the BBN group's strong beliefs about how it should perform in relation to the rest of the network. In particular, the BBN team tried to enforce the distinctions between network layers. An explanation of the IMP design decisions by team members John McQuillan and David Walden articulated their belief that the subnet should be isolated from any potential interference from the host computers—in other words, that the communications and host layers should be separate:

A layering of functions, a hierarchy of control, is essential in a complex network environment. For efficiency, nodes [IMPs] must control subnetwork resources, and Hosts must control Host resources. For reliability, the basic subnetwork environment must be under the effective control of the node program. . . . For maintainability, the fundamental message processing program should be node software, which can be changed under central control and much more simply than all Host programs. (McQuillan and Walden 1977, p. 282)

In the BBN vision, the IMP subnet was to be autonomous. Hosts would be isolated from failures in the subnet unless a host's own local IMP were disabled. IMPs would not depend on hosts for any computing resources or information, and the functioning of an IMP would not be impaired if its local host went down. Nor would the people at the host sites be able to interfere with the operation of an IMP in any way. Heart was particularly concerned that inquisitive graduate students would want to experiment with their local IMP, especially since computers of any sort were still a scarce commodity at most sites. He was anxious to make the IMPs self-contained, so that it would be hard for students to tamper with them (Hafner and Lyon 1996, pp. 121, 156). As practiced by Heart and his group, the technique of layering became a way to manage social relations as well as to reduce technical complexity. Designing the subnet to operate independent of the hosts made the network more robust, eased the technical task of the BBN team, and allowed the team to maintain control over the design and operation of IMPs.

The BBN group took several steps to make the operation of the IMPs depend as little as possible on the hosts, on other IMPs, or on human operators. Rather than counting on the network to function reliably, IMPs always checked for lost or duplicate packets, and each IMP tested periodically for dead lines, failures in neighboring IMPs, non-functioning hosts, or destinations made unreachable by intermediate IMP or line failures. The need for human intervention in the subnet was minimized by "ruggedizing" the IMP hardware. ("Ruggedizing," a common procedure for suppliers of military computers, entailed protecting the machine against temperature changes, vibration, radio interference, and power surges.) The team built into the IMP capabilities for remote monitoring and control that allowed BBN staff members to run diagnostic procedures or to reload software on an IMP without making field visits or relying on local operators. The IMP also was designed to recover from its own failures. An IMP that "went down" as a result of a power failure would restart automatically when power returned. Each IMP checked periodically to see if its basic operating program had been damaged; if so, it would request a neighboring IMP to send a copy of the program to replace the corrupted version. If the malfunctioning IMP was unable to reload the new copy, it would automatically shut itself down to protect the network from any destructive behavior the damaged software might cause. By anticipating and solving such day-to-day maintenance problems, the BBN team gave a practical form to the ideal of a distributed network.

As work on the communications layer proceeded, Roberts made new demands on the IMP, based on his evolving view of how the network would be used. Though the original plan had been to connect each IMP to a single computer, by 1971 it was apparent that some sites would need to connect multiple computers; thus, BBN modified the design of the IMP to accommodate multiple hosts (Ornstein et al. 1972, p. 244). In 1971 (two years into the project), Roberts decided to make the network accessible to users whose sites did not have ARPA-NET hosts. He directed BBN to create a new version of the IMP, called a "terminal IMP" or a "TIP," that would interface directly to terminals rather than to hosts. Once connected to a TIP, a user at a terminal could access any host on the network. The TIP dramatically extended the community of potential ARPANET users, since it was no longer necessary for a site to have its own time sharing host. By 1973, half the sites using the ARPANET were accessing it through TIPs (Roberts 1973, pp. 1–22).

Operating the Network: Redefining Responsibilities
In September of 1969, representatives from Bolt, Beranek and New-man and Leonard Kleinrock's group installed the first IMP at UCLA. This event marked the beginning of the ARPANET's operation, though the "network" had only one node at this point. By the end of 1969, less than a year after BBN won its contract, the IMP team succeeded in installing and linking the four initial nodes at UCLA, SRI, UC Santa Barbara, and Utah. But although the ARPANET was able to transmit test messages among the various sites, much work was still needed before the network could provide a usable communication system.

Bolt, Beranek and Newman's contractual responsibilities included keeping the IMP subnet running, and Frank Heart's team soon found that operating an experimental distributed network posed its own challenges. BBN set up a Network Control Center in 1970, when the company's own ARPANET node (the network's fifth) came online. At first the Network Control Center simply monitored the IMPs, and it was manned "on a rather casual basis" by BBN personnel (McKenzie 1976, pp. 6–5). However, as people began using the network, its reliability became an issue, and complaints from users forced BBN to take network operations more seriously. IMP and line failures were more common than BBN had anticipated, and, since the effects of a fault in one location tended to propagate across the network, identifying the source of the problem could be difficult. When a network user encountered trouble, Heart (1990, pp. 5–36) explained, you had the problem of trying to figure out where in the country that trouble was, whether it was a distant host, or whether it was the host connection, or whether it was an IMP at the far end, or whether it was in a phone line. . . . people certainly did not anticipate at the beginning the amount of energy that was going to have to be spent on debugging and network analysis and trying to monitor the networks.[17]

By late 1970, Roberts was also urging sites to make more use of the year-old network. When sites began turning to BBN with questions, the Network Control Center (NCC) took on the role of providing network information as well as handling trouble reports (McKenzie 1990, p. 20).

The responsibility for providing user support fell to Alex McKenzie. McKenzie had worked at BBN since 1967, but he did not become involved in the ARPANET project until November 1970, when he returned from an extended vacation looking for a new assignment. Frank Heart asked McKenzie if he would be willing to learn enough about the IMP to answer questions from the ARPANET sites, so that the IMP team could get on with its development work. McKenzie took charge of the Network Control Center in 1971, and he responded to the increasing demands from users by expanding and redefining the NCC's role. Believing that BBN should provide "the reliability of the power company or the phone company," McKenzie (1990, p. 13), promoted a vision of the ARPANET as a "computer utility." Under his direction, the NCC acquired a full-time staff and began coordinating upgrades of IMP hardware and software. The NCC assumed responsibility for fixing all operational problems in the network, whether or not BBN's equipment was at fault. Its staff monitored the ARPANET constantly, recording when each IMP, line, or host went up or down and taking trouble reports from users. When NCC monitors detected a disruption of service, they used the IMP's diagnostic features to identify its cause. Malfunctions in remote IMPs could often be fixed from the NCC via the network, using the control functions that BBN had built into the IMPs.

The NCC also gave the BBN group additional knowledge of, and therefore control over, the telephone network. Heart et al. (1970, p. 565) commented: "From the outset, we viewed the ARPA Network as a systems engineering problem, including the portion of the system supplied by the common carriers." BBN developed such expertise in diagnosing network troubles that the NCC was often able to report line failures before the telephone companies detected them—much to the carriers' surprise and initial skepticism (Heart 1990, p. 34; Ornstein et al. 1972, p. 253).[18] More generally, by building and operating its own switching network ARPA was able to control the characteristics of the communications system in such areas as cost, connection setup time, error rates, and reliability—areas in which computer users who relied on dial-up connections had little say. In this way the ARPANET represented a significant step toward integrating computing and telecommunications systems.

By 1976, the Network Control Center was, according to McKenzie (1976, pp. 6–5), "the only accessible, responsive, continuously staffed organization in existence which was generally concerned with network performance as perceived by the user." The initial division between subnet and host layers had simplified the work of the network's designers; now the NCC allowed the network's users to ignore much of the operational complexity of the subnet and to view the entire communications layer as a black box operated by Bolt, Beranek and Newman. The NCC had become a managerial reinforcement of ARPA's layering scheme.

Defining Host Protocols: A Collaborative Process
While the team at Bolt, Beranek and Newman was working out the design and operation of the subnet, the Network Working Group, led by Stephen Crocker at UCLA, began working on the protocol that would control the host interactions. Members of the NWG were excited to have the chance to explore fundamental computing issues of inter-process communications, but they were also daunted by the lack of prior work in this area, by the complexities of the software design, and by the need to coordinate the needs and interests of so many host sites. Having begun with some ambitious ideas, they soon realized that most sites were unwilling to make major changes to their hosts. They decided that the host protocols would have to be simple. Lawrence Roberts, in his earlier work with Thomas Marill, had even argued against requiring a network-wide host protocol: "Since the motivation for the network is to overcome the problems of computer incompatibility without enforcing standardization, it would not do to require adherence to a standard protocol as a prerequisite of membership in the network." (Marill and Roberts 1966, p. 428) But Roberts's earlier experiment had linked only two computers. A network with dozens of hosts would clearly need some level of standardization to avoid chaos.

Dividing the host functions into layers offered one possibility for making the task more manageable. Alex McKenzie, who was a member of the Network Working Group, recalled: "We had a concept that layering had to be done, but exactly what the right way to do it was all pretty unclear to us." (McKenzie 1990, p. 8) The NWG's initial plan was to create two protocols: one that would allow users to work interactively on a computer at another site (a process known as "remote login") and one that would transfer files between computers. Both protocols would occupy the same layer in the network system. Roberts, however, noted that both the remote login protocol and the file transfer protocol would have to begin their operations by setting up a connection between two hosts, and he saw this as a needless duplication of effort. Meeting with the NWG in December of 1969, Lawrence Roberts told the group to rethink the host protocol.[19] Crocker recalled: "Larry made it abundantly clear that our first step was not big enough, and we went back to the drawing board." (quoted in Reynolds and Postel 1987)

Roberts suggested separating the host functions into two layers. The first, called the "host layer," would feature a general-purpose protocol to set up communications between a pair of hosts; the second, called the "applications layer," would specify protocols for network applications such as remote login or file transfer (Karp 1973, pp. 270–271). Having separate host and applications layers would simplify the host protocol and lessen the burden on the host system's programmers. Also, eliminating the need for each application to duplicate the work of setting up a host-to-host connection would make it easier to create applications programs, thereby encouraging people to add to the pool of network resources. The ARPANET model now had three layers, as shown in table 2.3.

The host-layer protocol, implemented by a piece of software called the Network Control Program (NCP), was responsible for setting up connections between hosts. When an application program had data to send over the network, it would call on the NCP, which would package the data into messages and send them to the local IMP. Incoming messages from the IMP would be collected by the NCP and passed on to the designated application. The NCP also made sure that hosts communicating over the network agreed on data formats.

The NCP's design was shaped by assumptions about social and power relations in the networking community. Members of the NWG kept in mind that each ARPANET site would have to implement the

Table 2.3
The three-layer model of the ARPANET.

Layer	Functions
Applications	Handles user activities, such as remote login and file transfer
Host	Initiates and maintains connections between pairs of host processes
Communications	Moves data through subnet using packet switching; ensures reliable transmission on host-IMP and IMP-IMP connections

NCP: that is, someone at each site would have to write a program for the local host computer that would carry out the actions specified by the NCP. Since the host sites were rather reluctant partners in the ARPANET development effort, the NCP was designed to be simple, so as to minimize the burden of creating this host software. In addition, members of the Network Working Group were aware that the ARPANET system was being superimposed on existing patterns of computer use at the various research sites. The NCP's designers were therefore careful to preserve local control over the hosts by making remote users subject to the same mechanisms for access control, accounting, and allocation of resources as local users (Carr, Crocker, and Cerf 1970, p. 591). Finally, the NWG tried to preserve the autonomy of the ARPANET's users, many of whom were independent-minded computer experts. NWG member Stephen Carr noted: "Restrictions concerning character sets, programming languages, etc., would not be tolerated and we avoided such restrictions." (ibid., p. 79) The NWG outlined a design for the NCP early in 1970; by August of 1971, the protocol had been implemented at all fifteen ARPANET sites.

With the host protocol in place, the Network Working Group could focus on providing applications. The services originally envisioned by ARPA, and the first to be put in place, were remote login and file transfer. Early in 1970, several NWG members devised an experimental remote login program called telnet (for "telecommunications network"), which became generally available in February of 1971 (Crocker et al. 1972, p. 273; Carr, Crocker, and Cerf 1970, p. 594). Telnet initially formed the basis for other services, including file transfer, but eventually the NWG created a separate file transfer protocol called ftp.

Electronic mail was added to the system later, coming into general use around 1972. Telnet, ftp, and other applications went through continual revision as NWG members used these services and suggested improvements. One key to the ARPANET's success was that, since NWG members used the network protocols in their own work, they had the incentive and the experience to create and improve new services. In the process of working out applications that could run on different types of host machines, NWG members also addressed long-standing compatibility issues by developing common formats for representing files and terminals. These common formats became general-purpose tools that aided users of both networked and non-networked computers (Crocker et al. 1972, p. 275).[20]

Managing Social Issues

Lawrence Roberts, who regarded building a sense of community among ARPA's researchers as both a means to facilitate network development and an end in itself, coordinated the ARPANET project through a variety of informal mechanisms aimed at creating and reinforcing common values and goals. He maintained personal contact with his contractors through frequent site visits, which enabled him to check on the progress of the system and to reinforce commitment to the project. ARPANET participants could also meet at the Information Processing Techniques Office's annual retreats for Principal Investigators, praised by Frank Heart (1990, p. 40) as "among the most interesting, useful meetings that ever took place in the technical community." Formal presentations by PIs, with critiques from their peers, gave IPTO directors an opportunity to assess the progress of their various programs, and the small size of the meetings (generally less than 50 people) was conducive to informal sharing of ideas. IPTO's assistant director, Barry Wessler, ran similar meetings for the graduate students working on the ARPANET (Norberg and O'Neill 1996, pp. 44–46; Taylor 1989). By bringing researchers from around the United States together to work on pressing technical problems of mutual interest, PI retreats and graduate student meetings helped the social networks of computer scientists to become national rather than merely local.

Good relations among ARPANET contractors did not just make professional life more pleasant; they were also important to the project's success. In some cases, initial technical deficiencies made the very functioning of the network depend on the cooperation of its users. For instance, the first version of the IMP software did not effectively control the flow of data through the network, so it was possible for the system to become overloaded and break down. This was "fixed" in the short term when the users agreed not to send data into the network too fast; their voluntary actions thus compensated for the shortcomings of the system. There were also many points at which it was necessary for the groups at BBN, NAC, and UCLA to coordinate their efforts. BBN, through its monitoring of the network, provided crucial data to the network analysts at NAC and UCLA. There was no other source for data on the performance characteristics of a large distributed network. Since the theoretical tools of network analysis and simulation were in their infancy, they had to be checked against operational data (Heart et al. 1970, p. 557). Conversely, the analysis and simulation done by NAC and UCLA could aid BBN's engineering work by predicting potential IMP failures before they appeared in the network. For instance, when Robert Kahn of BBN and the group at UCLA worked together on network simulations, they were able to demonstrate that IMPs would be prone to congestion under heavy loads.[21]

This collaborative work was fruitful and often rewarding for those involved, but it also revealed or exacerbated tensions within the community. Despite an ethos of collegiality, there was also a good deal of potential conflict among contractors. Bolt, Beranek and Newman was at the center of many disputes. The company had much in common with ARPA's academic research sites: it was oriented toward research, and its development groups tended to be small and informal (unlike those at many large computer companies). The main IMP team had only five members, and the IMP software was designed, programmed, and debugged by three programmers (Heart et al. 1970, p. 566). But BBN was also very much a business, with an eye toward future profits. BBN's ARPANET contract represented a chance to get an early start in the new business of networking (in fact, the company would gain considerable revenues in later years from selling network services). To preserve its strategic advantage in having designed the IMP, BBN tended to treat the IMP's technical details as trade secrets. In addition, Heart was worried that, if BBN shared too much information about the IMP, graduate students at the host sites would try to make unauthorized "improvements" to the IMP software and would wreak havoc on the system. One of the more heated conflicts within the ARPANET community arose when BBN refused to share the source code for the IMP programs with the other contractors, who protested that they needed to know how the IMPs were programmed in order to do their own work effectively. The authorities at ARPA eventually intervened and established that BBN had no legal right to withhold the source code and had to make it freely available (Kleinrock 1990, p. 12; Schelonka 1976, pp. 5–19).[22]

The BBN team often had disagreements with the more theoretical groups at UCLA and at the Network Analysis Corporation. Most of the IMP team's members were engineers, not theorists, and their formative experience was in using available technology to build systems that worked—a process that required a certain amount of pragmatic compromise. The IMP team's priority—and its contractual obligation—was to get a working system in place quickly, and the members of the team tended to have little patience for academics who wanted to spend more time exploring the theoretical behavior of the network. In addition, since the complexity of the system made responsibility for problems difficult to pin down, contractors tended to trust their own work and find fault with others. Howard Frank (1990, p. 22) described NAC as having an "adversarial" relationship with BBN over the issue of congestion, which BBN attributed to shortcomings of NAC's topology and which NAC blamed on BBN's routing scheme.[23] Leonard Kleinrock described UCLA's relationship with BBN as one of "guarded respect," adding that "BBN was not very happy with us showing up their faults and telling them to fix them" (Kleinrock 1990, pp. 25–26). And the researchers at UCLA and NAC, who were jointly responsible for analyzing network performance, did not always agree; Kleinrock (ibid., p. 24) described their relationship ambiguously as "competitive but cooperative."

These tensions reflected the fact that the various groups involved in the ARPANET project had conflicting priorities for allocating their own time and effort. For instance, when UCLA's analysis predicted that the performance of the subnet would deteriorate under future heavy loads, Kleinrock urged the IMP group to revise the software right away, whereas the BBN team preferred to get the basic system installed and functioning before making improvements. Similarly, Stephen Crocker recalled that the NWG's first meeting with BBN in February of 1969 was awkward because of the different status and priorities of the two groups:

I don't think any of us were prepared for that meeting. The BBN folks, led by Frank Heart, Bob Kahn, Severo Ornstein and Will Crowther, found themselves talking to a crew of graduate students they hadn't anticipated. And we found ourselves talking to people whose first concern was how to get bits to flow quickly and reliably but hadn't—of course—spent any time considering the thirty or forty layers of protocol above the link level. (Reynolds and Postel 1987)

Within BBN, there was tension between Alex McKenzie's group at the Network Control Center, whose priority was to keep the system up and running reliably, and the IMP developers, who wanted to understand what was behind network malfunctions so as to prevent recurrences. When an IMP failed, the development team would often keep it out of commission for several hours while they debugged it, rather than immediately restoring it to service. Heart (1990) commented that the IMP developers came under increasing pressure as the network expanded and became more heavily used: "People began to depend upon it. And that was a problem, because that meant when you changed it, or it had problems, they all got mad. So that was a two-edged sword." The BBN group eventually resolved this conflict by developing new software tools that would make it possible to diagnose IMPs without keeping them out of service (Ornstein et al. 1972, p. 52).

It is testimony to the effectiveness of ARPA's management strategies that, despite these real conflicts of interest between contractors, the dominant paradigm remained one of collaboration. In a 1972 conference paper, representatives of the three main contractors—Howard Frank of NAC, Robert Kahn of BBN, and Leonard Kleinrock of UCLA—described how the ARPANET had provided a rare opportunity for collaboration across disciplines (Frank, Kahn, and Kleinrock 1972). They perceived their joint effort as something unique in computer science.[24] "Our approaches and philosophies," Frank et al. (ibid., p. 255) noted, "have often differed radically and, as a result, this has not been an easy or undisturbing process. On the other hand, we have found our collaboration to be extremely rewarding." Though they differed in their preferences for analysis, simulation, computerized optimization, or engineering experiment, after two years of experience they were willing to concede that "all of these methods are useful while none are all powerful." "The most valuable approach," they continued, "has been the simultaneous use of several of these tools." (ibid., p. 267)

If a lack of understanding among disciplines was considered the norm, ARPA's attempt to bring them together was all the more remarkable. Cultivating existing social networks, creating new management mechanisms to promote system-wide ties, and insisting on collaboration among groups all aided ARPA's social and technical integration of the system.

Preserving Informality: The Network Working Group

One of the most important mechanisms for pooling efforts and building consensus among the scattered sites was the Network Working Group. In assigning the NWG to create the host protocols, Lawrence Roberts had entrusted an important aspect of the system to a group of relatively inexperienced researchers. Vinton Cerf, then a graduate student at UCLA, described it as follows: "We were just rank amateurs, and we were expecting that some authority would finally come along and say, 'Here's how we are going to do it.' And nobody ever came along." (Cerf 1990, p. 110) Stephen Crocker recalled: "The first few meetings were quite tenuous. We had no official charter. Most of us were graduate students and we expected that a professional crew would show up eventually to take over the problems we were dealing with." (quoted in Reynolds and Postel 1987) The lack of established authorities and the newness of the field meant that the NWG's participants had to formulate technical problems and propose solutions on their own. "We were all feeling our way because there wasn't any body of current expertise or knowledge or anything," Alex McKenzie (1990, p. 8) recalled. When an outside observer—the RCA Service Company (1972, A-340)—asked why ARPA managers did not "take a more active role in defining the Host protocol,"

it was pointed out that it has been difficult to find the appropriate talent for this task. It is a curious blend of management and technical problems in that the decisions that would be made are relatively important and affect many implementations. It requires a fairly high level of systems programming experience [as well as] the ability to coordinate with a large number of people to realize a successful implementation.

At one point, Roberts, disappointed with the slow progress of the NWG, considered turning over the host protocols to a professional research team. In the end, however, he decided to stick with the NWG, in part because he sensed that the group increased the contractors' sense of involvement in and commitment to the network. As Carr, Crocker, and Cerf reported to a 1970 computing conference, the NWG provided a unique collaborative experience:

We have found that, in the process of connecting machines and operating systems together, a great deal of rapport has been established between personnel at the various network node sites. The resulting mixture of ideas, discussions, disagreements, and resolutions has been highly refreshing and beneficial to all involved, and we regard the human interaction as a valuable by-product of the main effort. (Carr, Crocker, and Cerf 1970, pp. 589–590)

In fact, the group's very lack of a firm blueprint for its actions gave it the flexibility it needed to balance technical and organizational issues. The NWG developed its own social mechanisms to ease the challenges it faced. Acting on a suggestion by Elmer Shapiro, Crocker proposed that technical proposals and minutes of meetings be distributed as a series of documents called Requests for Comments (RFCs). Another UCLA student, Jon Postel, took on the job of editing these documents. The RFCs were specifically designed to promote informal communication and the sharing of ideas in the absence of technical certainty or recognized authority. The NWG's "Documentation Conventions" stated:

The content of a NWG note may be any thought, suggestion, etc. related to the HOST software or other aspect of the network. . . . Philosophical positions without examples or other specifics, specific suggestions or implementation techniques without introductory or background explication, and explicit questions without any attempted answers are all acceptable. . . . These standards (or lack of them) are stated explicitly for two reasons. First, there is a tendency to view a written statement as *ipso facto* authoritative, and we hope to promote the exchange and discussion of considerably less than authoritative ideas. Second, there is a natural hesitancy to publish something unpolished, and we hope to ease this inhibition. (Crocker 1969)

During their first few years, the RFCs were, of necessity, distributed on paper; however, once the network was functional, the RFCs were kept online at the Stanford Research Institute's Network Information Center and were accessed through the ARPANET. Members of the Network Working Group would post new RFCs concurring with, criticizing, or elaborating on ideas presented in earlier RFCs, and an ongoing discussion developed. Eventually, after members had debated the issues through RFCs and at NWG meetings, a consensus would emerge on protocols and procedures, and this consensus was generally accepted by ARPA as official policy for the network. RFCs enabled the NWG to evolve formal standards informally.

Shaping the Political Environment
One potential source of tension that does not seem to have arisen within the ARPANET community was the involvement of university

researchers—many of them students—in a military project during the height of the Vietnam War. It helped that the network technology was not inherently destructive and had no an immediate defense application.[25] Perhaps the smoothness of the academic-military interaction merely reflects the self-selection of researchers who felt at ease in that situation. However, it is also true that IPTO managers were able to create an environment for their contractors that emphasized research rather than military objectives.

To a large extent, ARPA managers were able to shield their research projects from national politics, which sometimes conflicted with the agency's own priorities. ARPA's upper management became adept at buffering the agency's researchers from congressional scrutiny and from demands that they provide explicit military justifications for their work. In the late 1960s and the 1970s there were a number of US Representatives who believed that defense money should be spent only on projects closely tied to military missions (Norberg and O'Neill 1996, p. 36). They felt that the Department of Defense was becoming too involved in funding basic research, especially in view of the lesser sums provided by civilian agencies. In 1965, 23 percent of US government funding for university science came from the Department of Defense, only 13 percent from the National Science Foundation; in 1968, ARPA's budget alone was almost half that of the National Science Foundation (Johnson 1972, p. 335). During that year's Senate hearings on the defense budget, Senator Mike Mansfield of Montana challenged John S. Foster, the Director of Defense Research and Engineering, to explain why the Department of Defense should spend so much more on basic research than the NSF:

Senator Mansfield Is the answer partially due to the possibility that it is easier to get money for research and development in the Department of Defense than it is in any other department of the Government?

Dr. Foster No, sir; I believe the reason is deeper. I believe that the reason is that we are required to provide for national security. These amounts of money are required to provide assurance of an adequate technological capability. (US Congress 1968, p. 2305)

But, as Robert Taylor (1989, p. 27) privately admitted, the National Science Foundation's budget requests received closer scrutiny from Congress than ARPA's, since "the research pieces of the Department of Defense as compared to the development pieces of the Department of Defense were minuscule, whereas the National Science Foundation was *in toto* a research organization."

Though ARPA unquestionably played an important role in advancing basic computer research in the United States, the agency was careful to present Congress with pragmatic economic or security reasons for all its projects. It often characterized the ARPANET as an administrative tool for the military rather than as an experiment in computer science. For instance, in 1969 ARPA director Eberhardt Rechtin promised Congress that the ARPANET "could make a factor of 10 to 100 difference in effective computer capacity per dollar among the users" (US Congress 1969, p. 809). Two years later, ARPA's new director, Stephen Lukasik, cited "logistics data bases, force levels, and various sorts of personnel files" as military information sources that would benefit from access to the ARPANET (US Congress 1971, p. 6520). Once the ARPANET was up and running, Lukasik (1973, p. 10) reported to Congress that the Air Force had found the ARPANET "twelve times faster and cheaper than other alternatives for logistic message traffic."[26] Lawrence Roberts stressed expected cost savings in his public statements about the project (Roberts and Wessler 1970; Roberts 1973). Privately, Roberts observed that, when it came to classifying ARPA projects as either research or development, "we put projects in whatever category was useful, and I moved projects back and forth depending on how it was selling in Congress" (Roberts 1989). Roberts (ibid.) described his relationship with the US Congress as basically defensive: "We knew what routes would not work with Congress. So Congress clearly provided direction, even though it was more by stamping on you now and then. . . . We were not stamped on very often. I carefully constructed ARPA budgets so that we would avoid those problems."

Although ARPA's concern for defense applications and cost savings was genuine enough, the agency's disavowal of basic research was more rhetorical than real. John Foster, whose position as Director of Defense Research and Engineering included overseeing ARPA during the creation of the ARPANET, was a master of this rhetoric. Foster delivered this assurance to the Senate during its 1968 budget hearings:

The research done in the Department of Defense is not done for the sake of research. Research is done to provide a technological base, the knowledge and trained people, and the weapons needed for national security. No one in DoD does research just for the sake of doing research. (US Congress 1968, p. 2308)

Taken at face value, this statement might have surprised IPTO's academic contractors, since the agency was at the same time assuring them of its support for basic research and graduate education. Many of IPTO's computer science projects were proposed by the researchers themselves, or were designed to allow researchers to continue work in areas they had explored independently. Of IPTO, Taylor (1989, pp. 10–11) said this:

> We were not constrained to fund something only because of its military relevance. . . . When I convinced Charlie Herzfeld, who was head of ARPA at the time, that I wanted to start the ARPANET, and he had to take money away from some other part of ARPA to get this thing off the ground, he didn't specifically ask me for a defense rationale.

Even if the resulting technologies eventually became part of the military command and control system, the defense rationale might come after the fact. Describing his interactions with ARPA in the 1970s, Leonard Kleinrock acknowledged: "Every time I wrote a proposal I had to show the relevance to the military's applications." But, he claimed, "It was not at all imposed on us": he and his colleagues would come up with their own ideas and then suggest military applications for the research.[27] Wesley Clark's view was that, though IPTO contracts always specified some deliverable for the military, "Essentially, they were funding research with fairly loosely defined objectives. And the idea was to help them, whenever they needed help, to justify the work you were doing with respect to their sponsors in turn, the Department of Defense in general." (Clark 1990)

Obviously, ARPA contractors did not have absolute intellectual freedom. Vinton Cerf (who became an IPTO program manager in the mid 1970s) commented in 1990 that, although Principal Investigators at universities acted as buffers between their graduate students and the Department of Defense, thus allowing students to focus on the research without necessarily having to confront its military implications, this only disguised and did not negate the fact that military imperatives drove the research (Cerf 1990, p. 38). This was especially true in the late 1970s and the 1980s, when ARPA began to increase its emphasis on defense applications. However, during the period during which the ARPANET was built, computer scientists *perceived* ARPA as able to provide research funding with few strings attached, and this perception made them more willing to participate in ARPA projects. The ARPA managers' skill at constructing an acceptable image of the ARPANET and similar projects for Congress ensured a continuation of liberal funding for the project and minimized outside scrutiny. In this way ARPA was able to generate support from both its political and its research constituencies.

Launching the System

By the end of 1971 most of the infrastructure for the ARPANET was in place. The fifteen original sites were all connected to the network, which had begun to expand beyond the ARPA community to include sites run by the Air Force and the National Bureau of Standards. But most sites on the network were only minimally involved in resource sharing: the ARPANET had not brought about the radical jump in productivity that had been anticipated. Though the hardware and software developed for the system represented a great technical achievement, the network as a whole could hardly be considered a success if no one used it.

According to Robert Kahn (1990): "The reality was that the machines that were connected to the net couldn't use it. I mean, you could move packets from one end to the other . . . but none of the host machines that were plugged in were yet configured to actually use the net." The obstacle was the enormous effort it took to connect a host to the subnet. Operators of a host system had to build a special-purpose hardware interface between their computer and its IMP, which could take from 6 to 12 months. They also needed to implement the host and network protocols, a job that required up to 12 man-months of programming, and they had to make these protocols work with the rest of the computer's operating system (RCA Service Company 1972, p. A-72). Finally, the system programmers had to make the applications they had developed for local use accessible over the network. "This was uncharted territory, absolutely uncharted territory," Kahn (1990) recalled. "And people needed some motivation to get it done."

Early in 1972, Kahn and Lawrence Roberts decided that a dramatic gesture was needed to galvanize the network community into making the final push to get their resources online. They arranged to demonstrate the ARPANET's capabilities at the First International Conference on Computer Communications, which was to be held that October in Washington. Kahn took charge of organizing preparations for the demonstration, urging software experts to create new applications or make existing programs accessible over the network (Roberts and Kahn 1972). In the spring of 1972, the ARPANET team at BBN began to report "considerable enthusiasm" from the ARPA research community and an increase in traffic over the network (Ornstein et al. 1972). By the time the First International Conference on Computer Communications opened, enough programs were ready to capture the attention of the crowds.

The thousand or so people who traveled to Washington for the ICCC were able to witness a remarkable technological feat. From a demonstration area containing dozens of computer terminals, attendees were able to use the ARPANET to access computers located hundreds or thousands of miles away; there was even a temporary link to Paris. Software on these computers allowed participants to try out meteorological models, an air traffic simulator, conferencing systems, a mathematics system, experimental databases, a system for displaying Chinese characters, a computerized chess player, Joseph Weizenbaum's psychiatrist program Eliza, and a variety of other applications (Roberts and Kahn 1972). The diverse terminals, computers, and programs, all operating successfully and responsively, some across considerable distances, made a powerful impression. Cerf (1990, p. 25) later described visiting engineers as having been "just as excited as little kids, because all these neat things were going on." Another observer recalled: "There was more than one person exclaiming, 'Wow! What is this thing?'" (Lynch and Rose 1993, p. 10) The trade journal *Electronics* (1972, p. 36), citing "the great interest in computer networks indicated by . . . the crowds in the Arpanet demonstration room," declared networks "clearly . . . the wave of the future."

The ARPANET contractors had reported on the progress of the developing network at various professional conferences, but the response to the 1972 demonstration suggests that their colleagues did not necessarily take these reports seriously until they saw the network in action. "It was the watershed event that made people suddenly realize that packet switching was a real technology," recalled Kahn (1990, p. 3). The sheer complexity of the system, Roberts (1978, p. 1309) believed, was enough to make engineers skeptical until they witnessed it for themselves:

It was difficult for many experienced professionals at that time to accept the fact that a collection of computers, wide-band circuits, and minicomputer switching nodes—pieces of equipment totaling well over a hundred—could all function together reliably, but the ARPANET demonstration lasted for three days and clearly displayed its reliable operation in public.

Cerf (1990, pp. 25–26) noted "a major change in attitude" among "diehard circuit switching people from the telephone industry."

Though these communications experts had, with some justification, been skeptical of the idea of packet switching, they were able to appreciate the significance of the ARPANET demonstration, and it would be only a few years before the telephone companies started planning packet switching networks of their own.

The ICCC demonstration marked a turning point in the use of the ARPANET. Packet traffic on the network, which had been growing by only a few percent per month, jumped by 67 percent in the month of the conference and maintained high growth rates afterward (Schelonka 1976, pp. 5–21). In addition, the enthusiastic response to the demonstration encouraged some ARPANET contractors to start the first commercial packet switching networks.[28] In 1972 a group of engineers left BBN to form their own company, Packet Communications, Inc., to market an ARPANET-like service. BBN quickly responded to this defection by launching its own network subsidiary, Telenet Communications Corporation, and Roberts left ARPA to become Telenet's president. Telenet was the first network to reach the market, initiating service to seven US cities in August 1975. These new networks began to offer the general public the kind of reliable, cost-efficient data communications that the ARPANET had provided for a select few.

The triumphant public debut of the ARPANET was the culmination of several years of intense work in which the IPTO community developed a vision of what a network should be and worked out the techniques that would make this vision a reality. Creating the ARPANET was a formidable task that presented a wide range of technical obstacles and conflicts of interest. In the face of these challenges, the success of the project depended on the ability of the system's builders to foster a collaborative social dynamic among contractors, maintain financial support from Congress, and reduce the technical complexity of the system through techniques such as layering.

These strategies had lasting implications. ARPA did not invent the idea of layering; however, the ARPANET's success popularized layering as a networking technique and made one particular version of layering a prominent model for builders of other networks. The ARPANET also influenced the design of computers by highlighting the difficulties that existing machines encountered in a networked environment and offering some solutions. The host system programmers had demonstrated how to redesign operating systems to incorporate communications functions, and experience with the ARPANET encouraged hardware designers to develop terminals that could be used with a variety of systems rather than just a single local computer (Crocker et al. 1972, p. 275; Roberts 1970; Ornstein et al. 1972, p. 246).

The community that formed around the ARPANET ensured that its techniques would be discussed in professional forums, taught in computer science departments, and implemented in commercial systems. ARPA encouraged its contractors to publish their findings and provided funding for them to present papers at conferences. Detailed accounts of the ARPANET in the professional computer journals disseminated its techniques and legitimized packet switching as a reliable and economic alternative for data communications (Roberts 1988, p. 149).[29] Leonard Kleinrock's work became the basic reference in queuing theory for computer networks, and a number of graduate students that Kleinrock and others had supported based their later careers on expertise they had acquired while working on the ARPANET.[30] ARPA also encouraged its contractors to turn their ARPANET experience to commercial uses, as Lawrence Roberts had done with Telenet. ARPA's funding of Principal Investigators, its careful cultivation of graduate students, and its insistence that all contractors take part in the network project ensured that personnel at the major US computing research centers were committed to and experienced with the ARPANET technology. The ARPANET would train a whole generation of American computer scientists to understand, use, and advocate its new networking techniques.

3

"The Most Neglected Element": Users Transform the ARPANET

In light of the popularity of the Internet in the 1990s, we might expect that the ARPANET's first users would have quickly embraced the new technology. In practice, however, users did not move their research activities onto the network automatically or easily, and the results of such efforts were uneven. A number of diverse groups did make productive use of the ARPANET in the early 1970s, but other potential users were excluded or discouraged from using it, and many of ARPA's original predictions about how the network would benefit its users turned out to be wrong. The fact that the network became so successful is not something to be taken for granted, but rather something to be explained.

Historians have begun to call attention to the role of users in determining the features and ultimate success of a technology.[1] Typically, users are portrayed as consumers acting through the market, choosing one product or service over another. Occasionally, they are portrayed as concerned citizens pressing for regulations (e.g., safety standards). In any case, it is generally assumed that users become involved only after a technology has already been developed. But the ARPANET's ultimate "consumers"—the researchers who were to use it in their work—were directly involved in its development. During the ARPANET's first decade of operation, fundamental changes in hardware, software, configuration, and applications were initiated by users or were made in response to users' complaints or suggestions. It was, arguably, these activities that accounted for the perceived success of the system by ensuring that the ARPANET provided the types of services that users actually wanted. The ARPANET provides an instructive example of the variety of active roles users can play in shaping a new technology, and of the sometimes surprising results of their involvement.

"By No Means Complete or Perfect": The ARPANET as Experienced by Early Users

Many commentators on the popularized Internet of the 1990s have celebrated the advent of "cyberspace," the virtual realm in which people interact with computers and with other computer users.[2] Cyberspace provides an opportunity for individuals to create and explore imaginary environments, to experiment with different identities, and to establish new forms of community. Computer networks provide access to cyberspace, which appears as a welcoming, even playful environment in which newcomers receive instruction and encouragement from their fellow users.

The conditions encountered by the ARPANET users of the early 1970s stand in stark contrast to this rosy picture. Using the network and its host computers was difficult, the support systems were inadequate, and there was little opportunity to interact with other users. Michael Hart, one of the few early users from outside the field of computer science, later recalled that there was little on the net in the 1970s to attract users who weren't "computer geeks":

You have to realize how FEW people were on the Net before the '80s. . . . There just weren't enough to support a conversation on any but the most geeky or the most general topics. . . . It was boring, unless you could "see" down the cables to the rest of the world . . . and into the future. (Michael S. Hart, email to author, 28 March 1997)

There *was* a sense of community among many of the ARPANET's users, but it predated the network and was based on their shared backgrounds, interests, and offline experiences. One challenge in making the ARPANET user friendly lay in translating activities that build community—sharing of information, support, recreation—to the network environment. In taking these steps for the first time, early users of the ARPANET laid the groundwork for future virtual communities.

The road to becoming an active ARPANET user was long and hard. The first challenge for any potential user was getting access to the network. In order for a site to get an ARPANET connection, someone there had to have a research contract with ARPA (or with another government agency approved by ARPA). A prospective network member who was not being funded by ARPA had to pay the cost of setting up their node, estimated in 1972 to be somewhere between $55,000 and $107,000 (RCA Service Company 1972, p. A-72).[3] Once a site was approved, ARPA had to order a new IMP or TIP from Bolt, Beranek and Newman, direct the Network Analysis Corporation to reconfigure the network to include the new node, and arrange with AT&T for a telephone link between the new node and the rest of the ARPANET (ibid., p. A-81). The new host site would be responsible for providing the hardware and software for the host-IMP interface and for implementing the host protocol, NCP, on its computer(s)—a task that might represent a year's work for a programmer. In short, adding a new site to the network was complicated and costly. A prospective site's access was limited by its ability to pay, by the need to belong to an ARPA-affiliated research group, and by the need to have expert programmers available to create and maintain the host software.

Once a site was connected to the ARPANET, though, access controls were much looser. In theory, access within each site was to be limited to individuals doing work for ARPA. In practice, few sites tried to enforce that policy. Once a university or a company had connected a computer to the network, anyone with an account on that computer (or access to a friend's account) could use network applications such as email and ftp simply by executing the proper commands. Often, sites even included these unofficial users in the listings they submitted to the NIC "white pages," an online directory of ARPANET users (McKenzie 1997). Few system administrators tried to add access restrictions to the network commands. According to BBN's ARPANET Completion Report, "despite a deeply ingrained government and Defense Department worry about unauthorized use of government facilities, it was possible to build the ARPANET without complex administrative control over access or complex login procedures or complex accounting of exactly who was using the net for what" (Heart et al. 1978, p. III-111). BBN argued that this relaxed access policy made the system simpler and thus contributed to its quick and successful completion.

Many members of the ARPANET community suspected that ARPA managers were aware that unsanctioned users were on the network and did not object. Unauthorized users who contributed improvements to the system may even have received tacit encouragement from ARPA. In the early years the ARPANET was underutilized, and ARPA had little reason to discourage users or activities that might make the network more popular. Increased use of the network would also make it easier for ARPA's computer scientists to evaluate the system's performance. In fact, a recreational mailing list for "science fiction lovers" was apparently allowed to operate over the ARPANET on the ground that it generated significant amounts of traffic and therefore provided an opportunity to observe the network's behavior under load (McKenzie 1997). Another unofficial but tolerated activity was Michael Hart's Project Gutenberg, an effort to make historically significant documents available over the network. Hart, who was not an ARPA researcher but who had acquired an account at the University of Illinois, began by posting the Declaration of Independence on his site's computer in December of 1971; Project Gutenberg was still in operation on the Internet 25 years later.

Once on the network, users theoretically had access to some of the most advanced computer systems in the United States; however, using those remote systems could be difficult, impractical, or unappealing.[4] For one thing, new sites were provided with only scattered and incomplete resources to get them started. A 1972 report by an outside consultant stated: "The network user, new and established, is probably the most neglected element within the present development atmosphere. The mechanisms for assisting and encouraging new members are relatively informal or nonexistent." (RCA Service Company 1972, p. 9) ARPA provided an initial briefing to prospective members. Each site was given some printed documentation, including protocol specifications, a Resource Notebook containing descriptions of resources available at various sites, and a directory of participants (ibid., p. A-79). Some additional information was available online at the Network Information Center, located at the Stanford Research Institute. Beyond that, new sites had to find help where they could. Typically, they turned to BBN or to more experienced host sites for advice. ARPA did not provide in-depth training, and there was no single source to which users could turn for help in setting up network operations and locating resources (ibid., p. 29).

Just finding out what was available on the ARPANET could be difficult. The network search tools that Internet and World Wide Web users would later take for granted did not yet exist. Many sites did not provide complete or up-to-date information for the Resource Notebook, nor did sites generally offer online consultation about their resources, so users had to contact the sites offline to find out what services might be available (RCA Service Company 1972, p. A-21). Throughout the ARPANET's existence, its managers struggled to get host administrators to provide adequate information about their computer resources, technical configurations, and users.[5] It is not clear whether this was the case because paperwork was a low priority for computer scientists, or because the information was difficult for them to pin down (since site configurations were continually changing), or because they did not want to make this information available (perhaps fearing a loss of local control). What is clear is that the difficulty of learning about host resources was a major obstacle for new users. And there was no guarantee that host machines or resources, once located, would continue to be available: a research site might amass data for a particular project and then remove it when the project was completed, or it might temporarily take its machines off the network without notifying remote users (ibid., p. A-76).

There seemed to be general agreement among users that the Network Information Center, which was supposed to provide network information and a means for users to interact, was not working as planned. The Network Information Center did have some successes. Notably, the Network Working Group's protocol developers used the NIC's text editing and bulletin board systems to prepare, distribute, and store Requests For Comments. RFCs proved to be a very effective way for a large group to participate in ongoing technical discussions— in large part because members of the NWG (in contrast with many host administrators) were highly motivated to make information available to their scattered collaborators.[6] But as a directory of network resources, the NIC fell short, both because sites failed to supply information on their resources and because many people found the software at the NIC difficult to use (RCA Service Company 1972, p. A-9). To help fill the void, in March of 1973 the Stanford Research Institute began publishing the *ARPANET News*, a newsletter that listed updated information on host resources. The *ARPANET News* reduced (but did not eliminate) the difficulty of locating resources (Hafner and Lyon 1996, p. 229).

Users who had managed to identify an attractive remote resource faced another obstacle in the lack of administrative mechanisms for arranging the remote use of computers. Many host administrators wanted to charge remote users for computer time, or at least to know who those users were. This meant that users had to contact someone at the remote site to set up a computer account, and if they were going to be charged for their usage they also had to obtain a purchase order from their local institution. These interactions almost always had to take place off line, since few sites were prepared to conduct such business over the network. In addition to the extra administrative burden on the user, it was difficult for many researchers to get approval to spend computing funds at other sites rather than at their institution's own computer center. And since all these arrangements had to be made before the researcher ever got access to the remote resource, a potential user had to weigh a definite cost against an unknown benefit.

Connections from user sites to the network were not always satisfactory. Users at Wright-Patterson Air Force Base, which used a TIP to access the ARPANET, found this arrangement inadequate, since they had to use noisy telephone links to get to the TIP and since phone calls going through the base's switchboard were limited to 5 minutes' duration (Lycos 1975, pp. 161–162). Other TIP users may have experienced similar connection problems. A 1975 report by the US Geological Service, which had set up a conferencing system using the ARPANET, noted that "a major drawback of the early system was the unreliability of the experimental computer network we were using. Access was limited, and frequent hardware failures made 'real' work all but impossible." (ibid., p. 59) As a result, the USGS switched from using the ARPANET to using the commercial networks Telenet and TYMNET (Turoff and Hiltz 1977, p. 60). That these sorts of difficulties were not necessarily the fault of the ARPANET itself must have been little comfort to users who found themselves unable to communicate over the network.

Another technical obstacle was incompatibility between computers at different sites. Many of the hosts were unique systems with their own command languages and data formats, and some required specialized hardware at the user end. And, except for the few sites that hoped to generate income from network users, there was no great incentive for host sites to adapt their systems for remote use by others; thus, users were often left to deal with incompatibilities as best they could. Compatibility problems proved much more difficult to resolve than anyone working on the ARPANET seems to have expected.[7]

A user who had surmounted all these obstacles still had to figure out how to operate the remote computer. The instructions provided for the ARPANET demonstration at the International Conference on Computer Communications in 1972 convey some idea of how complicated using the network and its computers could be. The instruction booklet, entitled Scenarios for Using the ARPANET at the ICCC (Anonymous 1972), began with a disclaimer that "the scenarios are by no means complete or perfect." It urged participants to "approach the ARPANET aggressively" and to "unhesitatingly call upon the ICCC

Special Project People for the advice and encouragement you are sure to need." The detailed instructions for each scenario illustrated the many steps a user had to go through in order to connect a terminal to a TIP, tell the TIP which host to connect to, instruct the host computer how to handle the type of terminal being used, log in to the host computer, and run the desired application. Those who were already expert in various computer systems and ready to "approach the ARPANET aggressively" might find the network easy and exciting to use; for others, mastering the ARPANET must have been an uphill struggle.

Where could the novice user turn for help? Since most of the software available on the ARPANET had been developed as part of some local research project rather than as a commercial product, instruction and support tended to rely on informal local interactions. According to Alex McKenzie (email to author, 26 March 1997):

I don't think that any of the hosts were all that easy to use if you weren't part of the computer community. Most of the hosts were operated by researchers and tended to change frequently. Although every site had substantial documentation, it tended to *not* be tutorial in nature. You learned to use a host by talking to the other users down the hall. With "the hall" extended to intercity distances, it wasn't even easy for computer scientists to learn how to use a remote system, much less for other communities to do so.

However, experience with the ARPANET caused managers of host systems to reevaluate how they provided user support. Because none of the sites had served remote users before joining the ARPANET, the typical site's modes of support—such as system updates posted on bulletin boards or face-to-face interactions between users and support staff—had implicitly relied on users' having physical access to its computer center. In response to requests for help from ARPANET users, some sites began supplementing or replacing older means of support with online documentation, system announcements sent to each user's terminal, the ability to query support staff by email, and/or methods for "linking" terminals so that the user and the remote system operator would see the same screen output and could work through a problem together (Heart et al. 1977, p. III-8). In addition, most sites provided telephone consultation (which was toll-free at the NIC), and ARPA asked each site to designate a system specialist who would be available to answer users' questions.

Host sites' attitudes toward adopting these new methods for helping network users varied widely, depending on whether they saw outside users as a source of income or as a drain on local users' resources. Computer users at UCLA complained that they were not welcomed at other host sites: "Computer operations managers at other nodes may feel that incoming traffic is disruptive, less important than their own needs, or that UCLA's use of the net should be shunted to slack hours." (Brinton 1971, p. 65) On the other hand, sites that wanted to attract remote customers so as to generate income for their computer centers were eager to adopt new support techniques. For remote users of these sites, the network might indeed seem to be a welcoming place. But a significant number of the ARPANET's users were not satisfied with the services they were offered, and they began to take matters into their own hands.

Improving the System: User Activism

The ARPANET created an environment of both frustration and opportunity for its users. Using the network could be difficult, but a person with skill and determination (and there were many of these in the ARPANET community) could devise new applications with few restrictions. Thus, users had both the incentive and the ability to experiment with the system to make it better meet their needs. In some cases users built new hardware or software for the network, or asked ARPA or BBN to do so. In other instances, users improvised new ways of using the existing infrastructure. Users also began to organize to press for more support from ARPA—an activity that exposed tensions between segments of the ARPANET community. Three aspects of the system that users' experiments affected noticeably were terminal interfaces, connection paths between hosts, and applications protocols.

Terminal Interface Systems

ARPANET users had mixed relations with Bolt, Beranek and Newman, the contractor responsible for building and operating the network's infrastructure. The BBN team took great pains to respond to trouble reports and keep the network in constant operation, and they made a number of improvements to the interface message processor. However, they were more reluctant to respond to demands for new features and services, especially when these threatened to increase their own management tasks.

The BBN team made a number of changes from the original IMP specification as sites expressed their desire to use the IMPs in new ways (McKenzie 1997). The system was originally designed to have one host computer per IMP; however, users at UCLA—the very first site on the ARPANET—wanted to attach two computers to their IMP. To accommodate UCLA and other sites with multiple computers, Lawrence Roberts directed BBN to modify the IMP to handle more than one host. This paved the way for further innovations. Since a site's host computers were not always in the same location, multi-host IMPs tended to require longer connections to some of the hosts; since longer connections were more likely to suffer from line errors, BBN had to add error-checking procedures to the host-IMP interface. Users at UC Santa Barbara pushed the capabilities of the host-IMP interface to the limit when they requested a five-mile connection between a new host and their IMP so as to avoid having to install a second IMP. To accommodate this demand, BBN came up with a new interface, called VDH for "very distant host," that provided even more error checking.

The biggest innovation in the node design was to provide a way to connect terminals directly to the ARPANET, rather than expecting all terminals to be connected through host computers. Roberts decided in 1971 that such a terminal interface was needed in order to allow sites without host computers (such as ARPA itself) to connect to the network. This new type of node was called the terminal IMP (abbreviated "TIP").

The TIP greatly increased the numbers of sites and users that could access the ARPANET. However, many users were dissatisfied with the TIP interface, which represented a compromise between BBN's need for simplicity (in order to create a reliable system quickly) and the various requirements of terminal users. The most common complaint was that the TIP would handle only terminals for interactive computers ("asynchronous" terminals), while many people wanted to use terminals designed for batch processing computers ("synchronous" terminals). Some users also wanted to be able to program the TIPs to perform customized functions, such as reading files from magnetic tapes; they urged BBN to add new features to the TIP, or to allow users to modify their TIPs. Frank Heart, the head of the IMP group, commented: "Unfortunately, but perhaps not surprisingly, the limited goal and absolute restriction on user programming created considerable unhappiness in portions of the potential user community, and created considerable pressure for other 'better' terminal access techniques." (Heart et al. 1978, p. III-117)

From the perspective of Heart and the rest of the BBN team, these demands were not reasonable. The TIP was a small computer with no room for extra programs. The designers did not want to add additional disk storage, because this would make the TIP more liable to failure. Nor did the BBN team want to allow users to modify their TIPs: BBN was responsible for maintaining and upgrading the machines, a task that would be much harder if users were to make non-standard, possibly damaging changes. The BBN group felt that users simply did not understand how difficult it would be to provide user programming options, broad terminal support, and other special services.

Roberts had mandated some of the earlier changes in the IMP; however, perhaps swayed by Heart's arguments, he did not oblige BBN to make sweeping changes to the TIP. Instead, he gave financial support to users who wanted to develop alternative terminal interface systems.

Even before BBN had begun providing TIPs, some users had taken the initiative to build their own terminal interface machines. Such efforts continued after the TIP (with its perceived shortcomings) became available. In the first project of this kind, at the University of Illinois Center for Advanced Computation, W. J. Bouknight, G. R. Grossman, and D. M. Grothe designed a system they called the ARPA Network Terminal System (ANTS). The ANTS effort began in the summer of 1970, predating the introduction of the TIP. ANTS ran on a DEC PDP-11 minicomputer and was meant to provide an interface between an IMP and local terminals and modems. The system accommodated a wider range of terminals than the TIP, including the synchronous terminals typically used with batch machines. ANTS also allowed users to access other types of peripherals not supported by the TIP: users could read in data from punch cards, disks, or tapes for transmission across the network, and incoming data could be stored on these media or sent to local printers. Unlike the TIP, ANTS had disk storage, so users could keep files on their local ANTS system rather than on a distant host computer. ANTS also provided mechanisms for network access control and accounting.

The Illinois team put the ANTS system in place as soon as they received their IMP in April of 1971, and by September of 1971 ANTS was providing local terminal service. In the autumn of 1972 they added support for graphics terminals (Bouknight, Grossman, and Grothe 1973, pp. 73–74). The ANTS system gave the Illinois users a range of services that were not available from the TIP, and they considered it a great success. However, as BBN had warned, ANTS proved difficult to debug and maintain in the field. Furthermore, its complexity hindered transfer of the technology, and only a few sites ended up using it.

Another attempt to improve the ARPANET's terminal interface came from David Retz of the Speech Communications Research Lab at Santa Barbara. Retz had created a real-time data acquisition system for his ARPA-funded speech project, and he wanted to be able to send that data across the ARPANET for processing. The TIP, designed to handle commands typed from a terminal, was not equipped for such data transfers. Learning from the fate of ANTS, Retz and his colleagues decided to build a less ambitious terminal interface. Early in 1973 they began developing a system called ELF,[8] which also ran on a DEC PDP-11. Like ANTS, ELF allowed users to input and store files using local peripherals. ELF was more successful than ANTS, in part because it was simpler but also because its developers took advantage of the ARPANET to transfer the system to other sites. The ELF team would send the source code and binary files over the ARPANET to the target machine, perform remote debugging via the network, and keep in touch with remote ELF users through online release notes and bug reports (Retz and Schafer 1976, pp. 1012–1014). The initial ELF system was in experimental use at Santa Barbara in early 1974, and by 1976 about thirty ARPANET sites were using the ELF interface (ibid., p. 1007).

New Communications Paths
Members of the ARPANET community also found unexpected ways to use the network's communications links. The ARPANET was designed to connect distant computer centers, but users soon found a new application: sending data between computers at the same site. Local-area networks, which became ubiquitous in the 1980s, did not exist in the early 1970s; some manufacturers offered systems for networking their own line of computers, but there were no products for connecting different types of computers. Instead, users had to copy data onto tapes or other media and carry them from one local computer to the other. ARPA, concerned with the distribution of computing resources among different sites, had not focused on the networking of local computers, but users were very aware of the inconvenience of local data transfers.

When MIT's IMP was installed, in June of 1970, the ARPANET users there quickly realized that they could use the network to speed up communications between their local machines. No one had envisioned such a use of the ARPANET, and BBN's network monitors were puzzled when they started to notice that there was heavy traffic at the MIT IMP but not over MIT's outgoing lines. Eventually they realized that the MIT users had, in effect, turned their IMP into the hub of a local-area network (LAN). Soon other sites, among them Stanford University and the University of Southern California's Information Sciences Institute, began using the ARPANET as a LAN, and BBN itself began employing its IMP for local functions such as backups (McKenzie 1997). According to BBN's Frank Heart (1990), "the notion of using the IMP as a local connection was quite a surprise, to the extent that it became just common and had not been envisaged." By 1975 almost 30 percent of ARPANET traffic was intra-node (Heart et al. 1978, pp. III-77, III-91). A spontaneous innovation by users had contributed substantially to the use of the ARPANET and hence to its perceived value. Sites continued to use the ARPANET as a local-area network until "real" LANs based on Ethernet technology became available in the 1980s (McKenzie 1997).

Some users also created unusual or even illicit links between the ARPANET and other data communications systems. After the ARPANET was extended to England, physicists at the University of Illinois began using it to reach the Rutherford high-energy physics computer at Cambridge University. Rutherford had a separate connection to the Centre Européenne pour la Recherche Nucléaire (CERN) in Geneva through a European telecommunications carrier, but the carrier's regulations prohibited ARPA from using the Rutherford-CERN link to make connections from the United States to CERN. The Illinois physicists got around this restriction by using Rutherford as a dropoff point for files; by sending files or email between Illinois and CERN with a brief stop at Rutherford, they followed the letter of the law but were still able to create their own "virtual link" between the Illinois campus and the Geneva lab. Others used commercial networks to reach ARPANET sites. For example, John Day telecommuted from his home in Houston to his computer account at Illinois by setting up a connection from Houston to MIT through the commercial network Telenet, logging in to the MIT machine, and then going from MIT to Illinois through the ARPANET (John Day, telephone conversation with author, 11 April 1997). In both of these cases, the users needed to have accounts on machines at intermediate points (Rutherford, MIT) and needed to know how to use both networks, so making an inter-network connection required skill, resources, and motivation. Such "virtual internets" provide another illustration of how resource-ful users extended the capabilities of the ARPANET.

New Applications

Although many improvisations by users were encouraged or at least tolerated by ARPA, the agency did not always welcome users' attempts to steer the development of the system. This became painfully appar-ent when a group of user advocates tried to speed the development of upper-level protocols and applications for the ARPANET. In November of 1973, as enthusiasm for the network was beginning to grow among the ARPANET community, a group of systems developers who wanted to improve network services formed the Users Interest Working Group (USING). Members of this group began to critique the difficulty of using the network, and they lobbied ARPA to support the development of more and better applications. They also tried to create common tools for tasks such as accounting and editing; for example, they promoted a standard network editor called NetEd that was widely adopted.

Despite some initial support from ARPA, however, USING faced criticism when it tried to draw up a blueprint for the further develop-ment of the ARPANET's user services. Faced with organized action by users, the ARPA managers were evidently afraid that the network might slip out of their control. Members of USING were dissuaded from pushing their demands by ARPA program manager Craig Fields, who made it clear that the authority to make plans for the network lay with ARPA, not with USING. Early in 1974, Lawrence Roberts cut off funding for the development of upper-level protocols (Hafner and Lyon 1996, p. 230; John Day, telephone conversation with author, 11 April 1997).

The fate of USING revealed the limits of ARPA's generally non-hierarchical management approach. Individual users or research teams had tacit or explicit permission to add hardware and software to the system; ARPA even gave financial support for some of these experiments. However, users as a group had no say in the design decisions or funding priorities of the ARPANET project. The ARPA-NET experience is a reminder that the efforts of individuals to build virtual communities are constrained by the realities of money and power that support the infrastructure of cyberspace. ARPANET users continued to work for improved network applications, but after the demise of USING they focused on the more neutral activities of technical development and information sharing rather than on organized lobbying.

Rethinking the ARPANET's Purpose: Successes and Failures of Resource Sharing

When Lawrence Roberts described his original plan for the ARPANET, the goal he promoted was resource sharing: allowing individuals at different sites to share hardware, software, and data. The first published description of the ARPANET, co-authored by Roberts and his assistant Barry Wessler in 1970 and entitled "Computer Network Development to Achieve Resource Sharing," described the rationale for building the ARPANET as follows:

Currently, each computer center in the country is forced to recreate all of the software and data files it wishes to utilize. In many cases this involves complete reprogramming of software or reformatting the data files. This duplication is extremely costly. . . . With a successful network, the core problem of sharing resources would be severely reduced. (Roberts and Wessler 1970, p. 543)

The resource-sharing ideal was similar to the vision of a "computer utility" that was popular at the time.[9] Both models assumed that users would be accessing large, centralized machines (analogous to the generating plants of an electric power utility), with the network acting as a distribution system for computing power. The goal of resource sharing was partially fulfilled by the ARPANET: some sites did provide remote services to a significant number of ARPANET users. But, as we have seen, there were many obstacles to providing computer services in a way that was convenient for distant users. Many of the sites that succeeded had administrators who were strongly motivated to sell networked services; others had financial backing from ARPA to build computer systems that were specially adapted for use over the network. Lacking these incentives, most sites did not invest the effort needed to make their computers easy to use from afar. Overall, therefore, the practice of resource sharing on the ARPANET fell far short of Roberts's expectations.

Roberts had envisaged that the ARPANET would be used mainly to access time sharing computers, and his design specifications were aimed at supporting interactive computer use. As it turned out, however, only a few time sharing systems seem to have had significant numbers of remote users. One of these was the MULTICS operating system, created by MIT's project MAC in the 1960s, which featured a popular mathematics program called MACSYMA. By 1976 an estimated 15–20 percent of the MULTICS computer's load came from remote ARPANET users (Day 1997; Heart et al. 1977, pp. III-24–III-25). This percentage, though significant, was less than expected, in part because the managers of the MULTICS system did not welcome outside users.[10]

Another popular time sharing system was TENEX, which had been developed at Bolt, Beranek and Newman for the DEC PDP-10. BBN and USC's Information Sciences Institute were the main providers of TENEX service to remote ARPANET users; BBN offered its service on a commercial basis to ARPANET users, while the ISI was subsidized by the US government and handled jobs from government agencies (Roberts 1974, p. 47; see also Heart et al. 1977, pp. III-18–III-19). TENEX machines were particularly popular for text processing and artificial intelligence programming in the LISP language. Howard Frank's group at the Network Analysis Company used a remote TENEX machine to replace their old batch processing programs with interactive programs. Frank (1990) reported that access to the TENEX system sped up the planning of changes to the ARPANET topology: "We were able to create the same design in a day that was taking us two weeks to get before that."

Though ARPA favored time sharing, the ARPANET also offered the services of batch processing computers. Two sites provided access to large IBM batch processing machines. One was UC Santa Barbara, which had an IBM 360/75 that was mainly used for image processing. The other was UCLA, which had an IBM 360/91; the largest and fastest general-purpose computer on the network for many years, it served as ARPA's main "number cruncher" (Heart et al. 1977, pp. III-9, III-17).

UCLA had been entrepreneurial in putting the 360/91 on the network. In 1969 it had approached ARPA with a request for funding to build an interface between its computer and the ARPANET (UCLA Campus Computing Network 1974, p. 4). For UCLA's computer managers, who were financially responsible for an expensive and underutilized machine, the network connection would provide a way to sell extra computing capacity; for ARPA, the arrangement would provide its researchers with access to a high-end computer at competitive rates. Since the UCLA managers were eager to sell computer time, they made special efforts to meet the needs of remote users by providing expanded telephone support, a consultant dedicated to helping remote users, a Users' Manual that was stored on disk and could be printed out at a remote site, modifications in IBM's operating system to help keep local operators aware of the status of remote users' programs, and a "status" command that allowed users to follow the progress of their jobs through the system. UCLA's commitment to service attracted remote users, and within a few years ARPANET users were providing 10–20 percent of the UCLA computer center's income (Roberts 1974; Greenberger et al. 1973, p. 30).

Meanwhile, ARPA was building its own number-crunching machine: ILLIAC IV, a supercomputer with 64 parallel processors and a large memory. The ILLIAC IV project began at the University of Illinois, but was eventually transferred to NASA's Ames Research Center at Moffett Field in California. Since it had been built as an experiment in computer design rather than to serve a particular computing need, ILLIAC IV represented a solution looking for a problem. ARPA managers hoped that putting ILLIAC IV on the network would encourage researchers to find applications for the new machine. Remote ARPANET users at Rand, NASA, and other sites did use ILLIAC IV to run large-scale computations needed for climate simulations, signal processing, seismic research, and physics calculations (Heart et al. 1977, p. III-28); however, many of these projects had been created to find employment for ILLIAC IV, not to meet the pre-existing needs of users.

One large resource that had been explicitly designed for the ARPANET was the Datacomputer, a database system located at the Computer Corporation of America in Cambridge, Massachusetts. The Datacomputer had been designed by a team led by Thomas Marill, who had worked on early network experiments with Lawrence Roberts. Marill was a staunch believer in the resource sharing model. He held that in a networked environment host sites would tend to become specialized to take advantage of economies of scale, and that the availability of these specialized resources would eventually make the general-purpose computer obsolete. The Datacomputer was meant to be an example of the specialized resources that would dominate the future of computing (Heart et al. 1977, pp. III-29–III-30). It consisted of a DEC PDP-10 computer with a 3-trillion-bit storage device and with programs for storing, organizing, and retrieving very large amounts of data. To accommodate the needs of network users, the system provided data conversion facilities so that users could easily transfer data sets between different types of computers (Dorin and Eastlake 1976). The Datacomputer was heavily used by ARPA's seismic researchers, and the Argonne National Laboratory made it the repository of a climatological database. The Network Control Center stored statistical data on the performance of the IMP subnet on the Datacomputer, and MIT used it to store information it collected on ARPANET hosts (Heart et al. 1977, pp. III-35–III-36).

While some sites specialized in serving remote users, other sites became consumers of network-based services. Roberts pointed out that several sites, such as the University of Illinois Center for Advanced Computation (CAC), were able to dispense with local time sharing machines altogether and to contract for basic computing services from remote sites. Because the CAC's projects required diverse computer resources, the ARPANET was used to access several different types of machines—in particular, a Burroughs 6700 at San Diego. Starting in August of 1972, the CAC got over 90 percent of its computing services over the ARPANET, at about 40 percent of the cost of the former local operation (Sher 1974, pp. 56–57). At first, it should be noted, the Illinois researchers had not been enthusiastic about switching from local to remote computers. Michael Sher, the associate director of the CAC, acknowledged that the Illinois users had had to adjust to the unfamiliar practice of remote computing:

There was a great deal of skepticism among the center's programmers regarding their ability to develop systems and perform sophisticated applications programming over a network. Absolute control of computer resources, no matter what their quality, is normally not relinquished without significant reservations. (ibid., p. 57)

However, after being forced by economics to give up their local machine, the researchers found that networked computing had some unexpected advantages. Programmers could choose from diverse machines offering a range of services, including time sharing, fast calculation, and graphics routines. System developers were able to find remote users to test out experimental software (ibid., p. 58). In addition, with their newfound network expertise, the University of Illinois users were able to help their colleagues gain access to the network. As a result, Illinois researchers who were not funded by ARPA but who could suggest some defense-related application for their work were able to arrange to use the network for projects in economics, physics, and land use planning (John Day, telephone conversation with author, 11 April 1997).

Who Benefited?

Those who most readily benefited from the ARPANET were, not surprisingly, ARPA's computer scientists, who mainly used the network to trade files and information. Computer scientists had the expertise to use the system, and there were enough of them involved in the ARPANET project to form a community. And networking itself was a popular topic of research: one important ongoing activity was experimental research on topics such as switch design, protocols, and queuing theory. In addition to being a communications tool, then, the ARPANET was a source of empirical data and a test bed for new techniques.[11]

The ARPANET changed the way computer scientists worked and the types of projects that were feasible. Some collaborative projects, such as the development of the Common LISP programming language, would not have been possible without a means for extensive ongoing communication between many geographically separated groups (Sproull and Kiesler 1991, pp. 11, 32; Heart et al. 1977). In a 1986 *Science* article, several computer scientists noted: "The major lesson from the ARPANET experience is that information sharing is a key benefit of computer networking. Indeed it may be argued that many major advances in computer systems and artificial intelligence are the direct result of the enhanced collaboration made possible by ARPANET." (Jennings et al. 1986, p. 945) The National Institutes of Health sponsored a project at Stanford University to develop AI applications for medicine. This project supported AI studies at Rutgers, a medical diagnosis system at Pittsburgh, a database on eye disease at the University of Illinois, a distributed clinical database at the University of Hawaii, an effort at UCLA to model paranoid thought processes, and several activities at Stanford, including protein crystallography, an expert system for treating bacterial infections, and a program to aid chemists in determining molecular structures (Lycos 1975, pp. 193–197). Joshua Lederberg, a geneticist who had been one of the project's strongest advocates, described it as "one of the early 'collaboratories' enabled by the ARPANET" (message to Community Memory mailing list, 26 March 1997). In a 1978 article, Lederberg noted:

Such a resource offers scientists both a significant economic advantage in sharing expensive instrumentation and a greater opportunity to share ideas about their research. This is especially timely in computer science, a field whose intellectual and technological complexity tends to nurture relatively isolated research groups.

Computer scientists also used the ARPANET to share software and other files. Most collaborative projects involved the transfer of files containing documents or programs. A procedure for anonymous file transfer, implemented early on, made it possible to leave files in a "guest" account for anyone who wished to retrieve them. This allowed files to be exchanged informally, even without the originator's knowledge. The Stanford computer scientist Les Earnest recalled:

Another thing that happened a lot in the 1970s was benign theft of software. We didn't protect our files and found that both programs and data migrated around the net rather quickly, to the benefit of all. For example, I brought the first spelling checker into existence around 1966 but it wasn't picked up by anyone else, whereas the improved version (around 1971) quickly spread via ARPAnet throughout the world. (email to author, 28 March 1997)

In fact, in the 1970s a number of computer scientists had the impression that they and their colleagues were the *only* users of the ARPANET. David Farber, who was at Irvine, remembers only computer scientists being on the net (conversation with author, 22 February 1996). Les Earnest recalls that "nearly all ARPAnet participants in the early 1970s were computernicks. . . . I believe that there was very little academic or development activity outside of the realm of computer science" (email to author, 28 March 1997). A 1974 publication by UCLA describes the ARPANET as changing, between mid 1971 and mid 1974, "from a system programmer's experiment to an application programmer's tool" (UCLA Campus Computing Network 1974, p. 7)—hardly a move beyond the computer science community. According to these sources, computer experts dominated the network either because no one else was interested or because it was so difficult to use remote computers. Though there was some truth to this perception, use of the ARPANET did go beyond computing researchers.

Lawrence Roberts had announced in 1970 that the network would be used to support ARPA's researchers in behavioral science, climate dynamics, and seismology—researchers for whom computers were a tool, not a research focus (Roberts and Wessler 1970, p. 548). The man largely responsible for making this claim a reality was Stephen Lukasik, who in the early 1970s was both ARPA's director and the head of its seismology program. An early convert to the virtues of networking, Lukasik was (unlike Lawrence Roberts or Robert Taylor) involved in research areas besides computer science, and he sought out opportunities for ARPA's various contractors to work together. ARPA's use of the network for defense-oriented climate and seismic studies is a reminder that the ARPANET, though built by civilian research groups, was serving military needs from an early date.

ARPA's climate research program was one of the first to make serious use of the ARPANET. Predicting the weather has always been an important element in military planning; ideally, commanders would like to know seasonal weather conditions months in advance when planning attacks and invasions. Because of the chaotic nature of weather systems, however, such predictions are very difficult to make; to be feasible at all, they require very fast computers. Lukasik believed that climate modeling would be just the kind of data-intensive project that could provide a useful test of the ILLIAC IV supercomputer while also serving military needs. In the early 1970s he initiated a research program on global atmospheric circulation models that involved the Air Force, the Rand Corporation, the National Weather Bureau, the National Center for Atmospheric Research, Princeton University, and the Laboratory for Atmospheric Research at the University of Illinois.[12] In a typical use of the network, Illinois researchers would run large-scale hydrodynamic and meteorological simulations on the IBM 360/91 at UCLA, then send the output to the Information Sciences Institute (which had facilities for generating graphics), and finally send the graphics files back to Illinois, where the output would be displayed by local plotters. Climate researchers made extensive use of the ARPA-NET for remote job entry, file transfer, and interactive computing, and they reported that having access to the network allowed faster research and more efficient use of programmers' time (Sher 1974, p. 57; UCLA Campus Computing Network 1974, pp. 4–5; Heart et al. 1977, pp. III-17, III-54).

The seismology program—the other ARPA program that used the network in the early 1970s—was initially aimed at developing techniques for detecting tests of nuclear weapons in order to support a possible US-USSR treaty banning such tests. One sticking point to reaching agreement on such a treaty was the USSR's opposition to on-site inspection, the only known method of verification. Seismology seemed to offer a way out of this dilemma: if underground tests could be detected by seismic sensors located outside the USSR, it might be possible to provide verification of the treaty's terms without on-site inspection.

Detecting, locating, and identifying seismological events required large-scale data processing. To gather the raw data, hundreds of seismometers were arranged in arrays. ARPA had two of these arrays: the

Large Aperture Seismic Array in Montana and the NORSAR array in Norway. Monitors at each array would collect data and send it to ARPA's Seismic Data Analysis Center in Alexandria, Virginia, where analysts would use signal processing to look for patterns that would indicate likely seismic events and to characterize these events in terms of time, location, depth, magnitude, and probable cause (earthquake or explosion).[13] Before the ARPANET, magnetic tapes containing the seismic data had to be mailed to the SDAC; that made it impossible to examine seismic events in real time. Lukasik arranged to connect the Montana and Norway sites to the ARPANET (the latter through a satellite link), which allowed the SDAC to begin analyzing data within hours—rather than weeks—of an event. Since files of seismic data could be very large, the SDAC was also an ideal test user of the Datacomputer. By 1976, ARPA's seismic analysts had stored 70 billion bits of seismic data in the Datacomputer (Heart et al. 1977, p. III-35).

In addition to ARPA's own researchers, other government-funded scientists experimented with using the ARPANET. Physicists at several universities used it to access powerful computers elsewhere, such as UCLA's IBM 360/91 (UCLA Campus Computing Network 1974, p. 6). Chemists were able to access a molecular mechanics system at UC San Diego and a computational chemistry project jointly run by Wright-Patterson Air Force Base and the University of Chicago, and geologists could participate in a conferencing system set up by the US Geological Service in 1973 (Lycos 1975, pp. 42, 156).

Some members of the armed services used the ARPANET to access ARPA's computer resources or participate in its research projects (Stephen Lukasik, telephone conversation with author, 1 May 1997). The Army used the network to collaborate with ARPA's ballistic missile program, and the Air Force participated in ARPA's online seismic research. The Navy used the network to access the ILLIAC IV for acoustic signal processing. Researchers at the Aeronautical Systems Division of Wright-Patterson Air Force Base used the network to collaborate with colleagues at the Argonne National Laboratory on mathematical and chemical research (Lycos 1975, pp. 156, 174).

But despite this scattering of applications and users, most of the ARPANET's capacity went unused in the early years. Computer researchers, who were supposed to be the network's primary beneficiaries, used only a few of its remote computers to any significant extent. Though ARPA's seismology and climatology projects made use of the network, most other non-computer-science groups used it only for small-scale experiments if at all. (There is no evidence that it was ever used by ARPA's behavioral scientists.) The hope that the AR-PANET could substitute for local computer resources was, in most cases, not fulfilled.

The Decline of the Ideal of Resource Sharing

The ARPANET had been designed to give researchers access to computer resources that were presumed to be scarce. Ironically, however, many sites rich in computing resources seemed to be looking in vain for users. As the 1970s progressed, the demand for remote resources actually fell. ARPA managers went out of their way to find projects that had use for large computing resources such as ILLIAC IV and the Datacomputer. Despite a number of productive experiments using remote hardware and software, most members of the ARPANET community were not using the network the way it was originally intended: resource sharing, in the sense of running programs at remote sites, did not become the ARPANET's major purpose. In a 1990 interview, Leonard Kleinrock recalled:

Originally, the network was supposed to provide resource sharing. . . . For example, you would log on to Utah to use their graphics capability there. At one time it was thought maybe you could import the software to your machine and run it locally. But the original concept was that you would do resource [sharing] through the network—that's not really what happened.

A number of factors may account for the demise of the ideal of resource sharing. Non-expert computer users wanting access to resources at other sites faced a daunting array of obstacles, as has already been noted. On the other hand, in an example of the "not invented here" syndrome, computer scientists who had created hardware or software at their own sites often were uninterested in using machines at other sites; and those who did want to use a remote program were more likely to copy it and run it on their local machine than to run it remotely on its "native" machine (David Farber, conversation with author, 22 February 1996).

Many had expected that the network would be used for "distributed computing." This meant dividing a computing task among two or more machines, each of which would run part of the program; the various parts of the program would interact over the network as necessary. Distributed computing was supposed to allow users to combine the capacities of various specialized machines. In practice, however, the ARPANET project saw little if any realization of distributed computing.[14] Aside from the fact that there were incompatibilities between computers, the technical vision of distributed computing did not mesh with the administrative reality of using computer resources. The goal was to divide a task among various computers according to which machines were available and best suited for the job; ideally, the user would never need to know the details of the division of labor. In practice, however, because each site exercised administrative control over its computers, a user had to set up an account on each computer that might be used, and had to pay for whatever computer time was actually used. This made using multiple hosts more of a headache than a benefit to the average user (Heart et al. 1977, p. III-84). The problem was never solved, as Heart (1990) noted:

When the network was originally built, Larry [Roberts] certainly had high in his set of goals the idea that different host sites would cooperatively use software at the other sites. There's a guy at host one, instead of having to reproduce the software on his computer, he could use the software over on somebody else's computer with the software in his computer. And that goal has, to this day, never been fully accomplished. . . . So that turned out not to be the main thing that was created by the ARPANET.[15]

In addition to these technical and administrative obstacles, the attitudes that underlay the goal of resource sharing—both those of ARPA personnel and those of computer professionals in general—were beginning to change as the 1970s progressed. As the computer industry matured and a wider range of high-performance computers became available, ARPA managers no longer felt it necessary for the agency to build its own large machines; this meant that finding ways to share large computer resources became less of a priority (McKenzie 1997). At the same time, an increasing number of scientists were turning to smaller computers to meet their research needs. Minicomputers became popular in laboratories because they were much less expensive than paying for time on a large computer, and because a scientist could have control over an entire computer and could adapt it to the specific requirements of his or her lab. Looking back in 1988, Lawrence Roberts concluded that resource sharing had made economic sense only in the days when most ARPA researchers were using mainframe computers. Since mainframes provided computing power in large, fixed amounts, it was sometimes difficult to match the size of local resources to the needs of local users, and thus it may have been more cost effective to obtain computer resources over a network. Once a site could give each researcher a mini- or microcomputer, however, a network that provided access to computer resources no longer offered an economic advantage (Roberts 1988, p. 158). These developments undermined the old "computer utility" view that the point of a network was to help users access large, centralized computing systems. If extensive use was to be made of the ARPANET, it would have to be for some other purpose.

Finding a "Smash Hit": Email

Had the ARPANET's only value been as a tool for resource sharing, the network might be remembered today as a minor failure rather than a spectacular success. But the network's users unexpectedly came up with a new focus for network activity: electronic mail.

Email (initially called "net notes" or simply "mail") made an inconspicuous entry onto the ARPANET scene. Since many time sharing systems provided ways for users to send messages to others on the same computer, personal electronic mail was already a familiar concept to many ARPANET users. By mid 1971, when most of the sites had their host protocols in place, several ARPANET sites had begun experimenting with ideas for simple programs that would transfer a message from one computer to another and place it in a designated "mailbox" file. At the Stanford Research Institute, for instance, Richard Watson proposed such a system in order to make it easier for the Network Information Center to collect and distribute information about ARPANET sites (Watson 1971).

The first working network mail program was created by Ray Tomlinson, a programmer at Bolt, Beranek and Newman. Tomlinson modified the mail program he had written for BBN's TENEX operating system to specify a host name as well as a user name in the mail address, and he modified another command so it would transfer mail files between machines. His programs were incorporated into subsequent versions of TENEX, so that other ARPANET sites with TENEX machines were able to take advantage of the email feature (Ray Tomlinson, email to author, 10 November 1997).

In 1972, the Network Working Group was working on the specification for the file transfer protocol, which would replace the use of telnet for file transfers. Several people suggested making an addition to the ftp standard that would support email transfer. This was done at a March 1973 meeting of the Network Working Group, and it became possible to send messages using ftp instead of Tomlinson's TENEX-specific command. The ftp-based method of mail transfer was used until the early 1980s, when the NWG developed a separate mail protocol. (See Postel 1982.) Various members of the ARPANET community also wrote mail-reading programs that presented the contents of a user's mailbox in an organized way. One of the first mail readers was created by Lawrence Roberts; the first to become widely popular was MSG, written in 1975 by John Vittal at BBN.

Email quickly became the network's most popular and influential service, surpassing all expectations. The *ARPANET Completion Report* called its use by researchers for collaborative work the "largest single impact" of the ARPANET, noting that, along with the ability to easily share files, email had "changed significantly the 'feel' of collaborative research with remote groups" (Heart et al. 1978, p. III-110). Systems administrators began to use email for more mundane tasks, such as reporting hardware and software problems (Michael Hart, email to author, 26 March 1997). Inventive students participating in the early 1970s counterculture were rumored to use email for transcontinental drug deals (Les Earnest, email to author, 26 March 1997). ARPANET users came to rely on email in their day-to-day activities, and before long email had eclipsed all other network applications in volume of traffic. The *Completion Report* called electronic mail a "smashing success" and predicted that it would "sweep the country" (Heart et al. 1978, pp. III-113–III-115).

Email had several advantages over postal mail and the telephone. It was virtually instantaneous, and it did not require the sender and receiver to be available at the same time. Email programs were fairly simple to use even for computer novices, and the email addresses of registered ARPANET users could easily be found through the NIC. Copies of a message could be sent to several addresses at once, and widely dispersed groups could use email to coordinate their activities.[16] As email became popular, Roberts began funding "email hosts" at USC's Information Sciences Institute and at Bolt, Beranek and Newman; these provided email addresses for TIP users who did not have their own accounts on an ARPANET host (McKenzie 1997).[17] And ARPANET hosts that were connected to other networks (such as Telenet) often provided services for transferring mail between the two networks, further extending the community of email users.

Within the ARPA office and in the wider Department of Defense community, the use of email was vigorously promoted by Roberts and by ARPA director Stephen Lukasik. Roberts began using email to correspond with his contractors, thus giving Principal Investigators additional motivation to start using it themselves. Roberts found that email helped him overcome obstacles of time, space, and social distinctions in managing ARPA's many computer contracts. Alex McKenzie (email to author, 9 November 1997) recalled: "Email suited Larry's all-hours-of-the-night work habits and the far-flung set of projects he was responsible for. . . . He also liked the ability to use email to 'go around' the PIs and communicate directly with lower level employees." Lukasik also saw email as a convenient way to communicate with his managers and contractors. Roberts (1988, p. 168) recalled: "Steve Lukasik decided it was a great thing, and he made everybody in ARPA use it. So all these managers of ballistic missile technology, who didn't know what a computer was, had to start using electronic mail."[18]

At ARPA's headquarters, the appeal of the network had nothing to do with computers but everything to do with access to power. When Lukasik first got on the network, the only other people using email were computer scientists in the Information Processing Techniques Office. Program managers from ARPA's other offices began to notice that the IPTO contractors seemed to do better in the budget process because they were on closer terms with Lukasik. Wanting that same access, the other program managers began using email also. "It wasn't a technical issue," according to Lukasik (telephone conversation with author, 1 May 1997); "it was a management issue." "The way to communicate with me," Lukasik recalled, "was through electronic mail, and so almost all the offices then got on the net, and then the Strategic Office understood its utility and the Tactical Office understood its utility and my old Nuclear Monitoring Office understood its utility. . . . Of course, one can argue that even without me, everyone would be on networks because that's the way you work today, but in fact, you know, I really worked on it." From ARPA email began to spread to the rest of the military, and by 1974 "hundreds" of military groups were using the ARPANET for email (ibid.).

The popularity of email was not foreseen by the ARPANET's planners. Roberts had not included electronic mail in the original blueprint for the network. In fact, in 1967 he had called the ability to send messages between users "not an important motivation for a network of scientific computers" (Roberts 1967b, p. 1). In creating the network's host software, the Network Working Group had focused on protocols for remote login and file transfer, not electronic mail. Frank Heart (1990, p. 32) recalled: "When the mail was being developed, nobody thought at the beginning it was going to be the smash hit that it was. People liked it, they thought it was nice, but nobody imagined that it was going to be the explosion of excitement and interest that it became." A draft of the *Completion Report* referred to email as "unplanned, unanticipated, and mostly unsupported" (Heart et al. 1977, p. III-67).

Yet the idea of electronic mail was not new. MIT's CTSS computer had had a message feature as early as 1965, and mail programs were common in the time sharing computers that followed (Heart et al. 1977, pp. III-70–III-71). According to BBN's John Day (telephone conversation with author, 11 April 1997), it was an obvious step for programmers to expand these mail services to the network: "The paradigm was, what do we have on the operating system and how can we provide that on the network?"

Why then was the popularity of email such a surprise? One answer is that it represented a radical shift in the ARPANET's identity and purpose. The rationale for building the network had focused on providing access to computers rather than to people. In justifying the need for a network, Roberts had compared the cost of using the network against the cost of sending computer data by other media, but he had not compared the cost of email against the costs of other means of communication. The paradigm of resource sharing may have blinded the ARPANET community to other potential uses of the network.[19]

It was not that ARPA's managers did not value computer-mediated interaction. Indeed, Taylor and Roberts had expressed their hopes that the ARPANET would help build a community of researchers. But Roberts and others had expected users to collaborate by sharing files and programs or by using the centralized bulletin boards at SRI's Network Information Center. To help the NIC fill this role, Douglas Engelbart had created NLS, an information resource that provided a sophisticated environment for creating databases and conducting online discussions. But NLS was unfamiliar and confusing to many people, especially since most remote users lacked the specialized interface hardware that the system was designed to use (Heart et al. 1977, pp. III-67–III-68). Consequently, ARPANET users did not rely much on NLS as a way to interact with other people (McKenzie 1997). Had NLS been easier to use, it might perhaps have become the preferred method of communication, rather than email.

Computer experts put the first email programs in place, but non-expert users also had a role in building this new capability. In particular, ARPA director Stephen Lukasik used his influence to make sure that new features were added to the mail readers, so that a network tool that had been designed for computer scientists would meet the needs of other users. In the first mail readers, invoking the mail command would produce a printout on the screen of all the messages in one's mailbox, with the most recent messages last. For computer scientists, this was an appropriate format. "Computer scientists used mail in an almost real-time situation," according to Lukasik; they would send messages back and forth in rapid succession, "almost like a conversation." They usually had no need to keep old messages, so they had to deal with only a few messages at a time. But, Lukasik notes, "as a *manager,* you need to keep email for some time." Lukasik often asked contractors and subordinates for numbers and details on their projects, and he needed to save the answers for future reference. For him, email was not a conversation but a way to gather information. Typically he would find himself with fifty or so old messages in his mailbox, and he would have to scroll through all of them to get to the new messages at the bottom (Lukasik, telephone conversation with author, 1 May 1997). On hearing Lukasik complain about this, Roberts wrote a new mail program that sorted incoming mail into folders and made it easy to selectively view or delete messages.

Email and mailing lists were crucial to creating and maintaining a feeling of community among ARPANET users. Mailing lists allowed users to send messages to a single list address (such as sf-lovers@host-name) at a host site where a list administrator maintained a database of list members. A program running on the host computer then automatically re-transmitted the message to each person in the database. This meant that an individual could communicate with a large group without having to send out numerous messages and without having to keep track of the addresses of all the members. Even more important, mailing lists allowed a virtual community to take on an identity that was more than the sum of the individuals who made it up. For example, a science fiction enthusiast did not need to be personally acquainted with others in order to join an online discussion of the latest popular story; the names of lists advertised the common interests of their members, and most lists were open to all who wished to participate. Mailing lists provided a way for people to "meet" and interact on the basis of shared interests, rather than relying on physical proximity or social networks.[20]

The ARPANET Remade

The ARPANET had been ushered into the public eye with a triumphant demonstration at the International Conference on Computer

Communications in the autumn of 1972. But the ARPANET was not a finished product in 1972, nor was its success certain. The early years of the network were full of confusion, false starts, and frustration. Nonetheless, through many individual choices (and some top-down pressure from ARPA), people began to use the ARPANET and to discover how it could best serve their needs. These users were responsible for transforming the ARPANET from an experimental system with limited appeal to an operational service whose existence could be justified and even celebrated.

In the process of using the network, the ARPANET community developed a new conception of what networking meant. Since the original view of the network planners was that "resources" meant massive, expensive pieces of hardware or huge databases, they did not anticipate that people would turn out to be the network's most valued resources. Network users challenged the initial assumptions, voting with their packets by sending a huge volume of electronic mail but making relatively little use of remote hardware and software. Through grassroots innovations and thousands of individual choices, the old idea of resource sharing that had propelled the ARPANET project forward was gradually replaced by the idea of the network as a means for bringing people together.

Email laid the groundwork for creating virtual communities through the network. Increasingly, people within and outside the ARPA community would come to see the ARPANET not as a computing system but rather as a communications system. Succeeding generations of networks inspired by the ARPANET would be designed from the start to act as communications media. By embracing email, ARPANET users gave the network a new purpose and initiated a significant change in the theory and practice of networking.

4

From ARPANET to Internet

Over the course of a decade, the ARPANET (a single network that connected a few dozen sites) would be transformed into the Internet (a system of many interconnected networks, capable of almost indefinite expansion). The Internet would far surpass the ARPANET in size and influence and would introduce a new set of techniques to computer networking. For all its later importance, however, the Internet was not part of ARPA's initial networking plans. The Internet represented a new approach to networking, and its creation was prompted by a series of unforeseen events.

Even as the ARPANET was being developed, a small group of ARPA contractors were already working on the next generation of network technology for the military. In the course of trying to resolve some dilemmas they encountered in these other networking projects, ARPA researchers Robert Kahn and Vinton Cerf began to consider how to interconnect dissimilar networks.

The Internet architecture[1] that Cerf and Kahn proposed was eventually used not only to build the Internet itself but also as a model for other networks. One reason for the widespread acceptance of this new approach was that a range of interest groups participated in the Internet's design, including network researchers from outside the United States. Shaped both by ARPA's military concerns and by the opinions of an international community of network experts, the Internet would depart in significant ways from the design of the ARPANET, resulting in a system that was not simply bigger but also more flexible and decentralized.

Although the design of the Internet came from the international computer research community, the actual implementation was done under the auspices of the US military. Operational branches of the military began using the ARPANET during the 1970s, and

Department of Defense agencies other than ARPA became involved (somewhat reluctantly) in managing the network. The Department of Defense would play many roles in the emergence of the Internet: funding research and development, transferring technology to operational forces, using its financial resources to shape the commercial market for network products, and exercising management control over the ARPANET and its community of users.

New Directions in Packet Switching

The high-profile demonstration of the ARPANET at the 1972 International Conference on Computer Communications symbolically marked the completion of the original network project. That autumn also brought a change of personnel in ARPA's Information Processing Techniques Office. Lawrence Roberts left ARPA at the end of the year to head Telenet, BBN's commercial spinoff of the ARPANET. Around the same time, Robert Kahn, the BBN researcher who had been largely responsible for organizing the ARPANET demonstration, joined IPTO as a program manager.[2]

Kahn had no plans to pursue internetworking at this point; in fact, he did not expect to head a network project. On joining ARPA he had been told that he would manage a program in flexible manufacturing; when this was canceled soon after his arrival, Kahn was forced to find alternative projects for his office to fund. Since his main expertise lay in computer networking, he decided to look for defense-related projects in that area.

Kahn initiated an array of projects in network security and digital speech transmission. He also took up some investigations, begun by Roberts, of the application of packet switching techniques to two media that previously had not seemed suitable for data communications: land-based radio and satellite radio (hereafter referred to as "radio" and "satellite" respectively). Like the ARPANET, the experiments with radio and satellite combined fundamental research in computer science with potential for military applications. Packet radio seemed like an ideal medium for military field operations, since radio terminals (unlike the telephones of the 1970s) could be mobile. Satellite could provide worldwide communications and could support data-intensive defense applications, such as seismic monitoring of nuclear weapons tests; it also seemed suitable for use on Navy ships. Ironically, however, the biggest impact of the packet radio and satellite projects came not from their use in military operations (which turned out to be limited, though both were used in the 1991 Gulf War) but from their unplanned contributions to local-area networking and internetworking.

Alohanet and Ethernet

Packet radio posed a new set of theoretical and engineering questions. Radio, as a broadcast medium, has capabilities and limitations quite different from those of the wired telephone network. In a broadcast system, each message transmitted is received by every host; the host to whom it is addressed accepts the message, while the others ignore it. There is no need to route messages between individual hosts, and a mobile radio host can move around within the area covered by the broadcast without disrupting its ability to receive messages.[3] It is also easy to send a message to many hosts at once. Before these benefits could be realized in a computer network, however, a basic theoretical question had to be answered. Two messages sent at the same time on the same broadcast channel will interfere with each other, so both will be reduced to gibberish. How could a radio network be designed to prevent or to recover from such collisions?

The first experiments that set out to answer this question began in 1970, when Lawrence Roberts was still head of IPTO. They were led by Norman Abramson of the University of Hawaii and were funded in part by the Navy and by ARPA. The Hawaii group wanted to explore packet switching as an alternative to costly dial-up telephone connections for accessing the university's computers. The ARPANET was using leased telephone lines for its links, but Hawaii's noisy telephone lines were ill suited for data transmission (Kahn 1975, p. 177; Kahn 1989, p. 13). Abramson's group decided to try packet radio as a potentially cheaper and better means of serving the university's computer users. In 1970, Roberts provided IPTO funding for Abramson's Alohanet, which would link radio terminals at the University of Hawaii's seven campuses and numerous research institutes to its main computer center near Honolulu (Abramson 1970, p. 281). Abramson used a computer interface modeled after the ARPANET's IMP, and his team received design support from ARPANET veterans at the Network Analysis Corporation and at UCLA.

The Alohanet design used two radio channels. On one channel, the computer interface (named Menehune[4]) would broadcast packets from the computer center to the user terminals. Since the Menehune would be the only machine transmitting on this channel, there would no interference. The other channel would carry all the traffic from the users' terminals to the computer center. How could all the users share a broadcast channel without interfering with one another?

The Alohanet solution was startling in its simplicity. The designers did not try to prevent collisions at all; they simply made sure the system could recover from collisions when they occurred. The method was called "random access" because access to the channel by different terminals was not scheduled or coordinated; each terminal transmitted on its own initiative whenever it had data to send. If two users happened to transmit packets at the same time, both packets would be garbled. The system relied on acknowledgments to keep data from being lost in such collisions. The Menehune acknowledged safe arrival of a packet. If the sending terminal did not receive such an acknowledgment (perhaps because the packet had been destroyed in a collision), the sender would re-transmit the packet. But if *two* terminals lost packets in a collision, what would prevent them from re-transmitting at the same moment and endlessly repeating the collision? The Aloha answer was to have the terminals wait before re-transmitting, each terminal choosing its waiting time at random from a specified range. In all probability, the two terminals would choose different times to re-transmit, and thus they would avoid a repeat collision (Binder et al. 1975, p. 206). Figure 4.1 illustrates packet transmission on a random-access channel.

The Aloha method, a significant advance in communications theory and practice, became a standard topic in computer science textbooks. It also provided inspiration to Robert Metcalfe, a graduate student at Harvard. Metcalfe had been drawn into the ARPANET development effort through a part-time job at MIT's Project MAC, and the experienced had inspired him to write his doctoral dissertation on packet switching networks. To satisfy his Harvard committee, he needed to find a theoretical aspect of networking on which to focus. Metcalfe happened to be friendly with Stephen Crocker, then an ARPA program manager. Crocker gave Metcalfe some papers describing the Aloha project. Intrigued, Metcalfe read the papers and came up with a mathematical model that would significantly improve the performance of the Aloha network. His key insight was that varying the re-transmission interval in response to traffic loads—waiting longer to re-transmit when traffic is heavy—could radically improve the throughput of such systems by cutting down the number of repeat collisions. Metcalfe

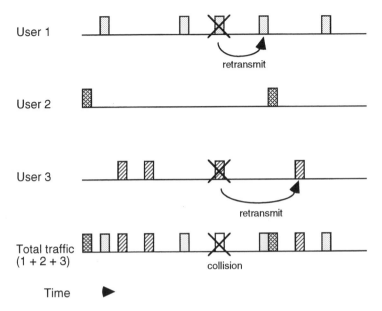

Figure 4.1
The Aloha technique. Packets from several users share a random-access channel. The different retransmit intervals keep the retransmitted packets from
colliding a second time. Adapted from Abramson 1970.

made the analysis of this phenomenon, called "exponential back-off,"
the heart of his 1973 dissertation, titled Packet Communication.

In 1972, while still working on his Ph.D., Metcalfe took a job at the
Xerox Palo Alto Research Center. PARC was a center of innovation;
its staff of talented researchers included Robert Taylor, who had initiated the ARPANET project. Taylor had left ARPA to become associate
director of the Computer Science Laboratory at PARC in 1970. In
1972 he was leading the development an innovative computer workstation called the Alto. A number of these workstations had been
deployed around PARC, and Metcalfe was asked to design a system to
connect them. Drawing on his dissertation work, Metcalfe created a
random-access broadcast system that was initially known as the Alto
Aloha network but was soon dubbed Ethernet (Thacker 1988, p. 274).
Ethernet used a cable rather than a radio channel as the transmission
medium. One advantage of using a cable was that it provided much
more bandwidth. Alohanet had transmitted thousands of bits per
second; Ethernet could carry millions per second. In combination with
Metcalfe's improved re-transmission algorithm, the use of cable made

Ethernet a fast and efficient way to transmit packets over short distances.

Recognizing the commercial potential of his invention, Metcalfe left Xerox to found a company called 3Com. In 1981 3Com announced an Ethernet product for workstations, and in 1982 it introduced a version for personal computers. For the first time, owners of small computers had an affordable networking option, and Ethernet quickly became a standard technique for local area networking. By the late 1990s, millions of LANs around the world were using Ethernet (Metcalfe 1996, p. xix). By providing the technical foundation for Ethernet, ARPA's first investment in packet radio had the unanticipated dividend of helping to spawn a huge commercial market for LAN systems.

Packet Radio and Satellite

Robert Kahn decided that ARPA should follow up on the Alohanet project by building a packet radio network of its own in the San Francisco Bay area. That system, called PRNET, consisted of a control station, several broadcast nodes (called repeaters), and a multitude of radio sets that could be attached to computers or terminals. The radio units were built by the Collins Radio Group of Rockwell International, the control stations were supplied by Bolt, Beranek and Newman, and the Stanford Research Institute was in charge of system integration and testing; the Network Analysis Corporation and the University of California at Los Angeles also participated (Kunzelman 1978, pp. 157, 160). PRNET went into experimental operation in 1975 with a single control station and four repeaters, as illustrated in figure 4.2.

Since radio was already used for command and control in the field, ARPA's packet radio program seemed more directly applicable to military operations than the ARPANET had been. Kahn was especially interested in using packet radio as a way to transmit voice for command and control. Packet switching could make voice transmission more efficient and could correct for the errors introduced by noisy radio channels. It would also make it harder for an enemy to eavesdrop on a conversation, since a message that had been digitized and split into packets would be unintelligible until it was reassembled and decoded at the receiving end (US Congress 1974, p. 135). The design of PRNET reflected its intended use in combat situations. To make the equipment easy to set up and use, Kahn specified omnidirectional antennas, so that users would not have to align them in the field, and he designed the system to keep track of the location and status of each

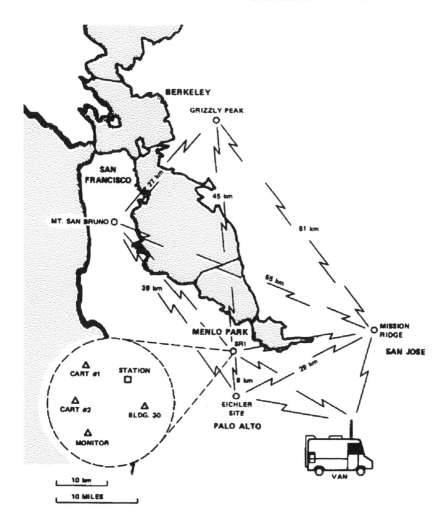

o = repeater

Figure 4.2
A map of the PRNET in 1977. The repeaters were located in elevated areas to increase their unobstructed transmission range. A radio-equipped van was used to test mobile communications. Source: Kahn et al. 1978.

component without human assistance (Kahn et al. 1978, p. 1478). Since the repeaters would operate unattended in remote outdoor areas, possibly in war zones, they were designed to be simple and rugged and to use minimal power. PRNET used various experimental data security techniques to prevent unauthorized access or tampering, and most control functions were located in the manned central station rather than in the more vulnerable repeaters. At the same time, PRNET made use of distributed control to survive an attack: the repeaters could take over routing without help from the control station (Kahn et al. 1978, p. 1417). Despite its careful design, however, PRNET was never developed to the point where it could be used in actual combat zones. ARPA managed to set up some test applications for the Army and the Air Force in which radio links were used to provide computer support for base operations, but the packet radio technology remained largely experimental (Cerf 1990).

ARPA ventured into another uncharted area with its packet satellite program. The development of packet switching networks in the 1960s and the 1970s had been paralleled by the development of satellite communications, which also had roots in the Cold War. In October of 1957, the USSR had launched the first artificial satellite, Sputnik I. Within a year, Sputnik had been joined in the sky by experimental satellites from the United States and Canada. In 1962, to encourage the use of satellites for peaceful purposes, President John F. Kennedy sent Congress a bill that became the Communications Satellite Act. This act supported the formation of a private corporation, Comsat, to provide commercial telecommunications service using satellites. At the same time, the United Nations was moving to create an organization to develop and operate a global communications satellite system that could be shared by all UN member countries. The International Telecommunications Satellite Organization (Intelsat) was established in August of 1964, with Comsat as the first US participant. Intelsat's first geosynchronous satellite was launched in April of 1965, and several generations of Intelsat satellites followed.

Satellites offered high bandwidth, and a single satellite could cover a wide area. Because of their high cost, however, satellite connections were rarely used for data transmission. Packet switching had the potential to make data communications via satellite economical. Kahn's immediate motivation for pursuing satellite network experiments was IPTO's seismic monitoring program (described in chapter 3 above), whose seismic sensors in Scandinavia generated voluminous streams of data that had to be transferred to the United States for analysis. Kahn decided that a packet satellite system would provide the most efficient way to do this, and he persuaded ARPA's seismic monitoring research office to shift its data transfer operations to a satellite link.

IPTO began using Intelsat I for experimental satellite links in 1973, connecting first the University of Hawaii and then University College in London to the ARPANET. In the autumn of 1975, Kahn began work on the Atlantic Packet Satellite Network (SATNET) project. Jointly sponsored by ARPA, the British Post Office, and the Norwegian Telecommunications Authority, the SATNET project was intended to support both network research and the transmission of seismic data for defense purposes (Jacobs et al. 1978).[5] In its initial configuration, SATNET linked four sites: one in Maryland, one in West Virginia, one in England, and one in Norway. The Norwegian site, run by the Norwegian Defense Research Establishment, was associated with the seismic monitoring program; the British site was operated by Peter Kirstein, a researcher at University College London who had arranged to participate in ARPA's network program. SATNET was a broadcast system; the four stations used a single radio channel to communicate with the satellite. The ground stations were connected to packet switches that were similar to the ARPANET IMPs but had been specially adapted to handle the high bandwidths and long transmission delays involved in satellite communications (ibid., p. 1461).

By the mid 1970s, then, ARPA was operating three separate experimental networks: ARPANET, PRNET, and SATNET. All these networks used packet switching, but they used it in distinctly different ways that optimized the technique for each particular medium. Kahn began to think about bringing these three networks together while he was struggling to develop PRNET into something more than an experiment.

PRNET connected a single computer center at SRI to a set of mobile radio units. Portable terminals could be attached to the mobile units, but portable host computers did not yet exist. To make the network useful, a way would have to be found to reach additional host computers. Kahn (1990) later described the situation as follows:

Partway through the first year of the program it became clear to me that we were going to have to have a plan for getting computer resources on the net. In 1973, mainframe computers were multi-million-dollar machines that required air-conditioned computer centers. You weren't going to connect them to a mobile, portable packet radio unit and carry it around. So my first question was "How am I going to link this packet radio system to any computational resources of interest?" Well, my answer was, "Let's link it to the ARPANET."

But linking PRNET and ARPANET was no simple proposition. The networks were technically incompatible: PRNET used broadcast, ARPANET point-to-point transmission; ARPANET guaranteed reliable transmission and sequencing of packets, PRNET did not; and packet sizes and transmission speeds differed between the two networks (Norberg and O'Neill 1996, p. 182). Moreover, Kahn (1989, p. 15) realized that he would eventually want to link the ARPANET to additional networks, such as SATNET, that used still other techniques. No one in the field of computing had ever attempted to connect such dissimilar systems, and there were no models from which to work.

As Kahn began thinking about ways to address the general problem of interconnecting heterogeneous networks, he set in motion what would become the Internet program.[6]

The Internet Program

In the spring of 1973, Kahn approached Vinton Cerf (then at Stanford University) with the idea of developing a system for internetworking. Cerf and Kahn had worked together on testing the first ARPANET node at UCLA, and Cerf had been one of the original designers of the ARPANET host protocol, so Kahn felt that Cerf was the right person to turn to for help. "It just took one session," he recalled (1990), "before the two of us were on the same wavelength as to what we needed to do. And he and I just jointly worked it out from there." The two collaborated on the initial design of a system that would link ARPA's various networks to form what would become known as the ARPA Internet. In the summer of 1973 they wrote a paper outlining the basic Internet[7] architecture (Cerf and Kahn 1974). Cerf received an ARPA contract to work out the detailed specifications of the system, and in 1976 he joined Kahn at ARPA and took over as program manager for the agency's various network projects.[8]

Starting from Kahn's original problem (how to access host computers from the packet radio network), Cerf and Kahn raised two basic questions (Cerf 1990). First, if the packet radio network were to provide reliable connections with the host computers, it would need a host protocol that could compensate for its error-prone transmission medium. What would that host protocol look like? Second, what kind of mechanism could provide an interface between two distinct networks such as PRNET and ARPANET? The answers they worked out would eventually become the basis for a set of internetworking techniques and for an experimental internet based on those techniques. But before they attempted to build such an internet, Cerf and Kahn sought out advice and opinions from the world's networking experts—a move that would significantly shape the resulting system.

An Inclusive Collaboration

Though Cerf and Kahn were the main architects of the Internet, they had a number of collaborators both from within the ARPANET group and from a growing international networking community. Among the members of the ARPA research community who were involved in designing the Internet were Yogan Dalal, Richard Karp, and Carl Sunshine (graduate students of Cerf's at Stanford); Stephen Crocker (who had worked on NCP as a graduate student and who was now at IPTO); Jon Postel of the Information Sciences Institute at USC; Robert Metcalfe of Xerox PARC; and Peter Kirstein of University College London (Cerf 1990, pp. 29, 33–34).

A number of computer researchers from outside the United States became involved in the project through the International Network Working Group, which had been formed at the 1972 International Conference on Computer Communications. The INWG brought together representatives from the world's major packet switching projects—the ARPANET, the British NPL network, and a French research network called Cyclades—and from various national telecommunications carriers who were planning packet switching networks of their own. The group soon affiliated itself with the International Federation for Information Processing (an association of technical societies that exchanged information and cooperated on the development of new technologies), thus adding to the INWG's visibility and legitimacy within the international computer science community.

Though the INWG had no formal authority to create international standards for computing, its members hoped to reach an informal agreement on internet standards so as to be able to interconnect their various systems (Curran and Cerf 1975, p. 8). Among the most active members from the United States were Franklin Kuo (who had worked on the Alohanet), Alex McKenzie of BBN, and Vint Cerf, who chaired the INWG from 1972 through 1976 (Cerf 1990). Cerf's involvement in the INWG allowed him to draw on the combined experience and expertise of this international networking community; it also encouraged him to expand the focus of the ARPA's internet program so that the proposed system would accommodate the various types of packet networks being built in Europe as well as ARPA's own networks.[9]

The ultimate conception and design of the Internet system would be shaped by the agendas of all these participants. In addition, the ARPA managers, who ultimately had to justify the program in military terms, wanted the system to support the complex requirements of the armed services. Writing in 1978, Cerf noted that computers were becoming ubiquitous in military equipment:

A fundamental premise of all current Command, Control and Communications (C3) research is that digital technology and computing systems will play a central role in the future. It is already apparent that computers are being employed in tactical as well as strategic military equipment. . . . To make this collection of computers, sensors, and databases useful, it is crucial that the components be able to intercommunicate. (Cerf 1979, p. 288)

For military purposes it was important that the Internet accommodate different types of networks, since it was expected that military communications systems would be optimized for a variety of service environments:

Ethernet ideas might serve well in garrison or aboard a ship. Packet radio concepts are crucial for local area mobile communication (e.g., land mobile, ground-air, ship-ship). ARPANET technology is appropriate for fixed installations such as in CONUS [the continental United States] or Europe. Finally, packet satellite supports wide geographic coverage while permitting efficient and dynamic allocation of transmission capacity as needed. The conclusion is that many different transmission technologies are needed for military operations and therefore, a sensible C3 system must incorporate a strategy for the interoperation of dissimilar computer communication networks. (Cerf 1979, pp. 288–289)

ARPA had started the packet radio and satellite programs to meet the military's perceived need for these different types of systems. The design of the Internet would also support this objective.

The members of the INWG were motivated by a common desire to enlarge the scope of their networks through interconnection, but they had divergent views on internet design principles. The most active French members came from the Cyclades project. Named after a group of islands in the Aegean Sea (since it connected isolated "islands" of computing), Cyclades was an experimental network project begun in 1972 with funding from the French government. Its architects, Louis Pouzin and Hubert Zimmerman, had very definite ideas about internetworking. In fact, Cyclades, unlike ARPANET, had been explicitly designed to facilitate internetworking; it could, for instance, handle varying address formats and varying levels of service (Pouzin 1975b, p. 416).

Cyclades was based on a very simple packet switching system. Rather than having the network maintain an ongoing connection between a pair of hosts, as the ARPANET did, Cyclades simply delivered individual packets (known as "datagrams"). Pouzin and Zimmerman argued that keeping network operations simple made it easier to build an internet. "The more sophisticated a network," according to Pouzin (1975b, p. 429), "the less likely it is going to interface properly with another. In particular, any function except sending packets is probably just specific enough not to work in conjunction with a neighbor." To keep the network's functions to a minimum, the French researchers argued, it was necessary for the host protocol to take on the primary responsibility for maintaining reliable connections. This went contrary to both the way BBN had designed the ARPANET and the way telecommunications carriers in France and elsewhere were planning to design their public data networks.[10] Perhaps anticipating opposition to their unconventional approach, the members of the Cyclades group were extremely vigorous in advocating their internetworking philosophy. Pouzin and Zimmerman were active in INWG. Another member of the Cyclades team, Gerard Lelann, worked in Cerf's lab at Stanford, where he was able to participate directly in the design of ARPA's internet system. According to Cerf (1990), the Cyclades group "had a lot to do with the early discussions of what the [host protocol] would look like."

England's National Physical Laboratory, which had pioneered packet switching techniques, was also involved in internetworking research. In 1971, a science and technology study group of the European Economic Community (precursor of the European Union) recommended the building of a multinational computer research network. The proposed European Informatics Network would help member countries share computer resources, would promote computer science research, and would provide a European test bed for networking techniques. Work began in 1973, and by 1976 the EIN was providing network service to ten countries.[11] Its British node was located at the National Physical Laboratory, and Derek Barber, who had worked on the original NPL network, led the development of EIN.

As part of the EIN experiment, researchers at the NPL set up a connection between their network and the EIN. They also made a trial connection between their network and the Experimental Packet Switching Service being offered by the British Post Office. In the course of these experiments, the NPL team confronted what they called the "basic dilemma" of internetworking: in order to get the most reliable and efficient service, it would be necessary to implement common host protocols on all the networks, but this would also require a substantial restructuring of existing network systems (Laws and Hathway 1978, p. 280). The NPL tried two approaches: for the EIN connection they translated between two different host protocols, while for the EPSS connection they used a common host protocol in both networks. Their experience confirmed that the translation approach was awkward and inefficient, and that establishing a standard host protocol would be the preferable way to build an internet (Laws and Hathway 1978, pp. 282–283).

Corporate researchers at Xerox PARC also played a significant part in designing the Internet. While Vinton Cerf and Robert Kahn were initiating the Internet program, Robert Metcalfe, David Boggs, and others at PARC were developing both the Ethernet local area network technology and a proprietary internet system called PARC Universal Packet ("Pup").[12] The initial Pup system was designed to connect several wide-area networks used by PARC (the ARPANET, PRNET, and the company's own leased-line network) and a number of LANs that used Ethernet (Boggs et al. 1979, p. 1). Drawing on ideas Metcalfe had presented in his 1973 dissertation, the Pup system provided a simple datagram service at the network level and relied on the hosts to provide reliable connections (ibid., pp. 3, 9). This approach to internetworking was similar to the Cyclades philosophy, but it arose from a local concern: the technical constraints of Ethernet. An Ethernet system has no "intelligence" inside the network; there is only a piece of cable connecting the computers, rather than a set of packet switching minicomputers as in the ARPANET. In an Ethernet system, therefore, the hosts must take most of the responsibility for running the network. This design was replicated in Pup. "An important feature of the Pup internet model," Boggs noted (ibid., p. 2), "is that the hosts *are* the internet." "Most hosts," he continued, "connect directly to a local network, rather than connecting to a network switch such as an IMP, so subtracting all the hosts would leave little more than wire." It is not surprising, therefore, that Metcalfe, when he joined the discussion on how to design the ARPA Internet, argued that the system should be based on the Pup approach of having simple network requirements and strong host protocols (McKenzie 1997).

Beyond Xerox PARC, which was oriented toward research rather than toward commercial production, there seems to have been no corporate participation in the design of the Internet. Few people outside the computer science community had even heard of the ARPANET in the early 1970s, and fewer still could have recognized that the Internet would someday become an important public and commercial technology. Like its predecessor, the Internet was designed, informally and with little fanfare, by a self-selected group of experts.

Designing the Internet

In June of 1973, Cerf organized a seminar at Stanford University to discuss the design of the proposed Internet and its host protocol, called the Transmission Control Protocol (TCP). All the interest groups mentioned above were represented at this meeting. As Cerf remarked in 1990, "TCP turned out to be the open protocol that everybody had a finger in at one time or another."

The seminar addressed the two questions originally raised by Cerf and by Robert Kahn: What was the best design for a universal host protocol that would work on unreliable networks such as the PRNET and not only on reliable ones such as the ARPANET? And how should the networks be attached to one another? Though these questions generated some debate, the participants were able to find enough common ground to define an approach on which most of them could agree. There was nothing inevitable about this agreement—as we will see in the next chapter, network design issues would become a source of international conflict by the mid 1970s. However, there seems to have been an emerging consensus among the computer researchers enlisted by ARPA on some basic principles.[13]

In answer to the first question, the group decided that TCP should have the responsibility for providing an orderly, error-free flow of data from host to host. Vinton Cerf, Gerard Lelann, and Robert Metcalfe collaborated closely on the specifications for TCP (Cerf 1990),[14] and thus the protocol reflected the design philosophies of Cyclades and Ethernet while deviating significantly from the approach that had been taken with the ARPANET. The ARPANET subnet was a very reliable communications system; the original host protocol, NCP, counted on this and did not have any error-recovery mechanisms. Cerf later pointed out that "even though the ARPANET was considered kind of a datagram-like system—because you put a label on the front [of each individual packet] and say 'here, deliver this'—underneath, inside the IMPs . . . things were delivered in sequence. And if they weren't in sequence there was something wrong." PRNET, on the other hand, might lose packets or deliver them out of sequence. "So we really needed a complete rethinking of the protocol suite," Kahn (1989, p. 19) recalled. TCP did much more than just set up a connection between two hosts: it verified the safe arrival of packets using acknowledgments, compensated for errors by re-transmitting lost or damaged packets, and controlled the rate of data flow between the hosts by limiting the number of packets in transit. All this made it feasible to provide reliable communications over a network as unreliable as PRNET. Cerf and Kahn planned for TCP to replace NCP as the ARPANET's host protocol and be the standard host protocol on every subsequent network built by ARPA.

Establishing a single universal host protocol was not the only possible approach to building an internet. One obvious alternative would have been to continue using different host protocols in different networks and create some mechanism for translating between them. This would have avoided the necessity of replacing existing host protocols, but Cerf and Kahn knew that such a design would not scale up gracefully: if the number of networks being connected were to grow large, the translation requirements would become unworkable. For Cerf and Kahn, the efficiency and flexibility of having a common protocol were worth the effort of converting the older system. Perhaps as important, the common protocol would create a particular type of experience for Internet users. According to Cerf (1990): "We wanted to have a common protocol and a common address space so that you couldn't tell, to first order, that you were actually talking through all these different kinds of nets. That was the principal target of the Internet protocols." Having to translate between different protocols would have emphasized the boundaries between networks, and the Internet's designers wanted the system to appear seamless. Indeed, they were so successful that today's Internet users probably do not even realize that their messages traverse more than one network.

To connect the networks physically, Cerf and Kahn proposed the creation of special host computers called "gateways." A gateway would be connected to two or more networks and would pass packets between them; all inter-network traffic would flow through these gateways. The gateways would maintain routing tables indicating how to send packets to each member network. Besides connecting networks, they would also help to accommodate differences between network systems by translating between the different local packet formats (Cerf 1979, p. 292).[15] Gateways buffered the local networks from having to know about the overall topology of the network. This made the system easier to scale up, because the local networks did not have to keep track of changes in the rest of the Internet; if a new network were added to the system, only the gateways would need to know.

This division of responsibility between local networks (which handled their own internal operations) and gateways (which routed messages between networks) was exemplified by the way the Internet handled addresses. ARPANET hosts had not needed addresses: packets were sent to a particular IMP, and it was assumed that a single host was connected there. If, instead, an entire network had been connected to that IMP, there would have been no way to specify which host on that network was supposed to receive the packet (Kahn 1989, p. 18).[16] The designers of the Internet had to devise a system of host addresses that would enable packets to be directed to a particular host on a particular network. They chose to create a hierarchical address system: one part of the address would specify the name of a network, while another part would give the name of an individual host within that network. The hierarchical address scheme facilitated the division of labor between gateways and local networks. Within each local network, the nodes would not have to know anything about non-local addressing or routing; they would simply send all packets with a non-local address to a gateway. Gateways would know how to route packets to any network, but they would not need to know the locations of host computers within that network (Cerf 1979, p. 297). This scheme kept the complexity of each part of the system manageable.

The system worked out by Cerf, Kahn, and their colleagues addressed the project's original requirements: it provided a protocol that would work over unreliable networks, and it solved the basic internetworking problems of routing and translating packet formats between networks. But some in the Internet group were critical of this initial design. Since the gateways used TCP, they ended up performing reliability functions (sequencing, error control, flow control) that were already being handled by the hosts; this made the gateways unnecessarily complex. The Xerox group compared this to "the complicated measures required to avoid deadlock conditions in the Arpanet—

conditions which are a direct consequence of attempting to provide reliable delivery of every packet" (Boggs et al. 1979, p. 8). They argued that the gateways should provide only a simple datagram service.

At an internetworking meeting at the University of Southern California in January of 1978, Vint Cerf, Jon Postel, and Danny Cohen discussed this issue and came up with a solution. They proposed splitting the TCP protocol into two separate parts: a host-to-host protocol (TCP) and an internetwork protocol (IP). The pair of protocols became known collectively as TCP/IP. IP would simply pass individual packets between machines (from host to packet switch, or between packet switches); TCP would be responsible for ordering these packets into reliable connections between pairs of hosts (Cohen 1978, p. 179). The idea of having a separate internet protocol was modeled in part on Xerox's Pup network, which was being developed around the same time (Cerf 1980, p. 11; Boggs et al. 1979, pp. 4–6).

With the new version of the Internet protocols, gateways could be simpler: they would run only IP, and they would no longer have to duplicate the host functions (now confined to TCP). The minimal functions required of the Internet Protocol also put fewer demands on member networks.[17] Introducing the new set of protocols, Cerf (1980, p. 10) wrote: "The Internet Protocol (IP) has been designed around the premise that few assumptions can be made about the type of service available from any given network." He also argued that the stripped-down functionality of IP would make military networks more robust and therefore more likely to "meet the requirements of operation under hostile conditions" (ibid., p. 11) He noted that, by accommodating diverse networks, the design would allow the armed forces to create specialized networks and also to introduce new technologies over time without major disruption of the system. "Thus," he noted elsewhere, "the problems of dealing with dissimilar tactical and strategic networks and with evolving computer communication network technology can be solved in a single stroke." (Cerf 1979, p. 289) But the version of TCP/IP that became standard in 1980 was more than a military product; it also reflected the ideas and interests of an international community of network researchers.

Initial Experiments
Designing the protocols was only the first step toward building the Internet. Putting the design into practice took several years; even in simplified form, the network protocols performed a complex set of

functions and were difficult to implement correctly in software. The initial version of TCP was specified in 1974. Bolt, Beranek and Newman had an implementation of TCP for the TENEX operating system completed in February of 1975, though it was not reported to be debugged until November. BBN also built its first experimental gateway in 1975, connecting an in-house research network to the ARPANET. Stanford University implemented TCP during 1975, and in November the BBN and Stanford groups set up an experimental TCP connection between their sites. The early tests revealed a number of deficiencies in the design, forcing the Internet group to revise the TCP specification (McKenzie 1991a).

The BBN group proposed testing TCP over the satellite network, and they began installing experimental gateways at BBN (in 1975), at University College London (in November of 1976), and at the Norwegian Defense Research Establishment (in June of 1977) (Travers 1991). As they conducted tests over these links, the Stanford and University College London researchers discovered that badly programmed implementations of TCP could drastically degrade the network's performance (Bennett and Hinchley 1978, p. 406). The protocol specification was only a "blueprint"; it was up to the host system's programmers to make a working version of the protocol—a lesson that would become painfully clear when the entire ARPANET community tried to adopt TCP. By late 1977, however, the test sites were ready to try out the new protocols, and ARPA demonstrated its first multi-network connection. Experimenters sent packets from a van on a California freeway through PRNET to an ARPANET gateway, then through the ARPANET to a SATNET gateway on the East Coast, over SATNET to Europe, and finally back through the ARPANET to California (figure 4.3).

For the computer scientists, the 1977 demonstration confirmed the feasibility of the Internet scheme. For ARPA, it was also a way of highlighting the military potential of the new technology. Cerf (1990) emphasized that "all of the demonstrations that we did had military counterparts," suggesting how connections between radio, satellite, and telephone networks could be used during wartime:

What we were simulating was a situation where somebody was in a mobile unit in the field, let's say in Europe, in the middle of some kind of action trying to communicate through a satellite network to the United States, and then going across the US to get to some strategic computing asset. . . . There were a number of such simulations or demonstrations like that, some of which

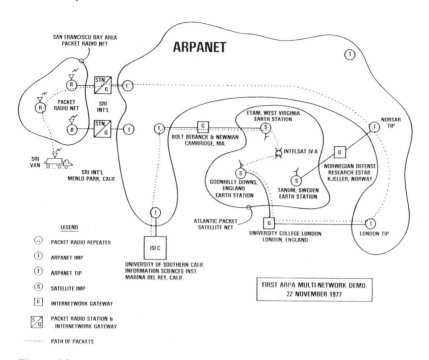

Figure 4.3
Diagram of 1977 Internet demonstration.

were extremely ambitious. They involved the Strategic Air Command at one point, where we put airborne packet radios in the field communicating with each other and to the ground, using the airborne systems to sew together fragments of Internet that had been segregated by a simulated nuclear attack.

The successful three-way interconnection of the ARPANET, PRNET, and SATNET represented the beginning of the Internet as an operational system. The design of the Internet made it possible for the networks to operate independently but still communicate, which benefited ARPA's experimental network projects. For instance, SAT-NET researchers could use the ARPANET to coordinate project personnel, monitor SATNET equipment, and generate test traffic; at the same time, SATNET remained a separate system from the ARPANET, which gave researchers the freedom to conduct possibly disruptive experiments on SATNET without disturbing ARPANET users (Jacobs et al. 1978, pp. 1462–1464). The ability of local networks to maintain their autonomy while participating in the Internet also made it easier to include networks from outside ARPA. After the demonstration, a number of new defense and research networks joined ARPA's evolving

Internet, including the Defense Communications Agency's Experimental Data Network, the Army's Fort Bragg packet radio network, various Ethernet LANs at Xerox PARC, an experimental packet radio network at BBN, the network of MIT's Laboratory of Computer Science, and the British Post Office's Experimental Packet Switching System (Cerf and Kirstein 1978, p. 302).

To encourage sites to adopt TCP, ARPA began funding implementations of it for various operating systems. In 1977, ARPA funded BBN to incorporate TCP/IP into the popular Unix operating system, and one of the system's creators, Bill Joy, added TCP/IP to the Berkeley version of Unix.[18] ARPA also funded implementations for IBM machines, for the DEC TOPS-20 system, and for other operating systems (McKenzie 1997).

The ARPANET, however, did not adopt TCP/IP immediately. ARPA managers encouraged host sites to implement the new protocol, but did not force them to do so. Most sites chose to continue using NCP: the old protocol was providing perfectly adequate service within the ARPANET, and researchers who were not actively involved in internetworking experiments had no immediate motivation to switch protocols. Implementing TCP was difficult; to make matters worse, the specification kept changing as the Internet team adopted new ideas and as experimental use revealed shortcomings in the design. It was not ARPA's research community, therefore, that pushed for the transition from the ARPANET to the Internet.

Military Involvement in the Internet

The impetus for adopting TCP/IP came from the operational branches of the military (the armed forces and the agencies that support their day-to-day operations). Not all commanders were eager to adopt ARPA's new networking techniques, and there was often a clash of cultures between the ARPANET's research and military communities. But a combination of circumstances caused the Defense Communications Agency, which provided communications services for the armed forces, to view the ARPANET as an important part of its own system-building plans. As the DCA began to depend on the ARPANET, its managers took an active role in guiding the system's technical evolution and eventually championed full adoption of the Internet protocols.

The operational defense agencies first became interested in the ARPANET as a model for replacing their existing networks with more advanced technology. The National Security Agency commissioned Bolt, Beranek and Newman to create two smaller versions of the ARPANET for the intelligence community,[19] and the Defense Communications Agency experimented with ARPANET technology as part of its plans to upgrade the WorldWide Military Command and Control Systems.

In 1962, during the Cuban Missile Crisis, President Kennedy had discovered that the US military did not have an effective worldwide communications system for command and control. To remedy this, the Department of Defense had initiated the WorldWide Military Command and Control Systems, which built on an earlier system devised for the Strategic Air Command.[20] WWMCCS consisted of a hierarchy of command and control centers around the world that were equipped with computer systems to gather data on the status of forces and to store war plans (Zraket 1990). But the initial communications system for WWMCCS, which used leased lines to connect the computers, was far from satisfactory. In fact, since the modems used were so slow, the personnel at the centers often found it quicker to put data on a tape and transport it on an airplane than to transfer it over the phone lines (Eric Elsam, telephone conversation with author, 22 July 1997).

The Defense Communications Agency was eager to try more advanced technology for its new WWMCCS network, called WIN. In 1972 the agency contracted with Bolt, Beranek and Newman for a three-IMP network called PWIN that was used to develop software and test operations for WIN (US Congress 1972, p. 822). After a successful demonstration of PWIN, the DCA built the operational WIN network. In addition to transferring techniques, hardware, and skilled personnel from the ARPANET into a new military project, the WIN project convinced a number of people at the DCA that packet switching represented the future of data communications.[21]

Soon after beginning the WIN network, the DCA took on a new and unexpected role as the ARPANET's operator. ARPA, as a small research agency, was not well suited to provide routine data communications services. Once the ARPANET had passed the experimental stage, ARPA began looking for a new operator. In 1972, ARPA and BBN began to consider transferring the ARPANET to another government agency or a commercial carrier, with the hope that it would grow into a nationwide public service (Ornstein et al. 1972, p. 253; McQuillan et al. 1972, p. 752).[22] After discussing the matter with the Federal Communications Commission and other agencies, ARPA's managers decided to find a commercial operator who would buy the network hardware from ARPA, receive an FCC license as a specialized common carrier, and supply communications services to the government and other customers (US Congress 1972, p. 822). A 1974 report that ARPA commissioned from Paul Baran concurred that moving network operations to competitive commercial suppliers would stimulate the US networking industry and make it easier for military and civilian users to share the use of the ARPANET (Kuo 1975, p. 13).[23]

Many members of the ARPANET community, including Robert Kahn, Lawrence Roberts, and Howard Frank, took part in the effort to find a new operator. AT&T, the largest telecommunications carrier in the United States, seemed the most likely candidate. Roberts and Frank met with AT&T managers to explain how the network could be scaled up for commercial use (Roberts 1978, p. 49; Frank 1990, pp. 26–27; Kleinrock 1990, p. 36). AT&T declined, perhaps because the packet switching business was too small and too different from conventional telecommunications to seem worth its while. In 1975, after lengthy discussions among Roberts, ARPA director George Heilmeier, and other DoD personnel, the ARPA managers decided to temporarily transfer operational responsibility for the ARPANET to the Defense Communications Agency. ARPA would continue to provide funding and technical direction, and access would be open to DoD users and to government contractors approved by the DCA. The agreement left the fate of the network after three years unresolved, since ARPA still hoped to find a home for the ARPANET outside the government; nonetheless, the DCA ended up operating the network well beyond the initial three-year period.

It may seem unusual that the operational branches of the military took so little heed of the ARPANET before 1975. But ARPA played an unusual role in the Department of Defense: ARPA's whole purpose was to pursue research projects that were far ahead of the contemporary state of the art and were not tied too closely to specific applications (such as weapons systems). This role freed ARPA to look beyond the immediate concerns of the armed forces, but it also meant that ARPA sometimes had to work hard to get other military agencies interested in its innovations. It was only after the DCA took over operation of the ARPANET that the network began to be used by the armed forces in any extensive way.

The ARPANET as a Defense System

After officially assuming control of the ARPANET on 1 July 1975, the Defense Communications Agency began to reorient the network away from its research origins and toward routine military operations. Military users made increasing use of the network now that they could arrange connections through the DCA. "It was their normal way," Kahn (1990, p. 40) observed; "they didn't have to deal with a research agency." This effort to transfer ARPA's network technology to the various commands was aided by a personnel change within the Information Processing Techniques Office. Army Colonel David Russell became director of IPTO a few months after the DCA took over the ARPANET, and he helped to accelerate contacts with the armed forces and to promote the ARPANET as a test bed for new computerized command and control systems (Kahn 1990, p. 38; Klass 1976, p. 63).[24] By 1976, the Air Force, the Navy, and the Army were all using the ARPANET to experiment with such systems.[25]

The DCA imposed its own style of management on the ARPANET. The DCA's ARPANET manager, Major Joseph Haughney, commented:

When the network was small, a decentralized management approach was established due to the nature of the network and the small community of users. This promoted flexibility and synergy in network use. Now that the network has grown to over 66 nodes and an estimated four to five thousand users, flexibility must be tempered with management control to prevent waste and misuse. (Haughney 1980a)

Under the new regime, prospective sites had to go through a more involved process to get access to the ARPANET; Bolt, Beranek and Newman and other contractors had to work through bureaucratic rather than informal channels; and there were more rules for what could or could not be done with the network (McKenzie 1997). The DCA was more serious than ARPA had been about preventing use of the network for "frivolous" activities, even if these activities did not disrupt network operations. For instance, in March of 1982 the DCA's new ARPANET manager, Major Glynn Parker, complained about an "email chain letter" that had been circulating on the network and threatened to cut off hosts whose users forwarded the letter (Parker

1982a). The DCA also wanted to cut down on the computer scientists' common practice of copying files across the network without their owners' explicit permission, which had become an accepted way for users to share the latest improvements in network software. The DCA expressed concern that government information might be inappropriately released to the general public or sold to industry, and it instituted a new policy that a file's owner had to give explicit consent before any copying could be done (Haughney 1981b).

As military use of the ARPANET grew, the DCA also tried to enforce the network's access policy, which many researchers felt had been more honored in the breach. In theory, the benign neglect of access controls that had allowed system administrators to turn a blind eye to unauthorized users would no longer be tolerated. Frequent reminders in the DCA's online newsletter that "all unauthorized use of the AR-PANET is prohibited"[26] suggest that local administrators were not quick to enforce this policy, however. Some host administrators did not even know who all their ARPANET users were, since their computers were not set up to control which users could access the network. Haughney (1981b) warned these administrators that they would have to start monitoring or restricting access to their machines:

If unauthorized users are found on the net because of a weak or nonexistent host access control mechanism, we will review the host's access mechanisms and request improvements. If the host refuses a review or refuses to make the suggested improvements, we will take action to terminate its network access. This is a club of last resort, but we will use it to protect other network users who have invested time and money to bring their controls up to par.

Haughney presented these measures as necessary to protect the military computer systems from malicious infiltration, stressing that the aim of the new access controls was "to ensure that we can verify proper resource utilization and prevent unauthorized penetrations" (1980a).

The DCA's heightened concern with network security was a response to wider trends in computing in the 1970s. In January of 1975, only a few months before the DCA assumed control of the ARPANET, the world's first personal computer was introduced in the United States. The Altair 8800 was made by a small company called Micro Instrumentation Telemetry Systems and advertised in the magazine *Popular Electronics*. It was primitive, and it was sold as a kit, but its price was astonishingly low: $379. The Altair 8800 was an instant hit with amateur computer enthusiasts, who place thousands of orders during the first few months it was advertised. Suddenly, a technology that had been restricted to authority figures in academia, business, and government was in the hands of teenage hobbyists. Members of a new "hacker" subculture quickly made improvements to the Altair and began devising more user-friendly machines, and by the late 1970s there was a thriving market for personal computers.

The spread of computer expertise to a much wider segment of the American population increased the risk that hackers would be able to break into restricted military systems on the ARPANET. Computer administrators had only to look to the telephone system for an example of the type of "unauthorized penetrations" Haughney was worried about. The 1970s saw the widespread use of "blue boxes"—devices that mimicked the control tones used by the telephone system—to fraudulently obtain free phone service (AT&T Bell Laboratories 1982, pp. 430, 432). A number of highly publicized incidents dramatized how pranksters known as "phone phreaks" used blue boxes to make free calls all over the world, often just for the challenge of mastering the telephone system. Phone phreaks came from the same world of young, undisciplined technophiles as computer hackers; for instance, before Steve Jobs and Stephen Wozniak started Apple, they had been in the business of making and selling blue boxes (Campbell-Kelly and Aspray 1996, p. 244). Haughney (1981a) warned ARPANET host managers that "the advent of lowcost, home computer systems has subjected the ARPANET to increased probing by computer freaks."

DCA managers were particularly concerned about the TIPs—the network nodes that allowed users to reach the ARPANET by dialing up from a terminal rather than having to go through a host computer. Initially, anyone with a terminal and the telephone number of the local TIP could use the ARPANET. To increase security, the DCA instituted a new system of logins and passwords to ensure that only authorized TIP users would have access to the network.

Another unforeseen set of circumstances spurred the DCA to become involved in the Internet effort. In 1976, DCA managers decided to procure from Western Union an upgraded data network to replace the outdated AUTODIN (Automatic Digital Network), a message switching network that the DCA had built for the military in the early 1960s. The new network, called AUTODIN II, was meant to replace AUTODIN, WIN, and the military sites on the ARPANET. AUTODIN II was slated to go into operation late in 1979 and would connect some 160 host computers and 1300 terminals (Kuo 1978, p. 309).[27] The DCA considered dismantling the ARPANET once

AUTODIN II had been constructed, but the agency eventually concluded that there was still a role in the Department of Defense for a research-oriented network. Disruptive experiments would clearly be out of place on an operational military network such as AUTODIN II, and they would be just as unwelcome on a commercial data network, which would be the main alternative for researchers if the ARPANET were dismantled. Therefore, the DCA planned to leave the research portion of the ARPANET intact and to set up gateways to connect it to AUTODIN II (Kuo 1978, p. 312).

The DCA's decision to create an internetwork link between the ARPANET and AUTODIN II meant that the agency suddenly had a need for ARPA's new Internet protocols. Kahn and Cerf had been actively promoting TCP/IP as a potential standard for DoD networks, and in 1980 the Office of the Secretary of Defense formally adopted the ARPA protocols—which were still somewhat experimental—as military standards (Kahn 1990; Parker and Cerf 1982; Cerf 1980, p. 11). As Kahn (1990) explained, TCP/IP was the only system available that could meet the DoD's needs:

We needed to switch over to the internet protocol because connections between multiple nets needed an internet protocol. . . . The sweep of events at the time was such that DoD really had to decide what guidance to give people who were connecting their computers to the net as newer sites came in. "What do we tell them?" So they finally decided to standardize [TCP/IP], because it was really the only game in town at that point.

It was these pragmatic considerations, rather than any demand from the research community, that drove the DoD to take the first decisive step toward making TCP the standard for ARPANET hosts.

By 1981 the armed services—which were being asked to pay for and use the new system—were complaining that AUTODIN II was too expensive and technically deficient. Don Latham, the Assistant Secretary of Defense for "C3I" (Command, Control, Communications, and Intelligence), asked the DCA to come up with an alternative, but the agency was not able to do so. Latham then appointed Colonel Heidi B. Heiden, who had been the Army's planning director for computer communications, to join the DCA and put together a team to come up with an alternative network design. The DoD did not immediately abandon the AUTODIN II effort; rather, it gave Western Union and Heiden's team six months to prepare their systems for evaluation by a DoD review board, which would choose one (Heidi Heiden, telephone conversation with author, 30 July 1997).

Heiden did not want to build new hardware, as the AUTODIN II group was doing. His plan was to use the DoD's existing packet switching networks—the ARPANET, WIN, and MINET (a version of the ARPANET used in Europe)—as the basis for a new Defense Data Network (Heiden and Duffield 1982; Harris et al. 1982). He wanted to use commercial technology wherever possible, to cut development costs and to give the DoD competing sources for its components. He believed that ARPA's Internet protocols would provide the best service, and he defended them against rival standards, such as the protocols that were then being developed by the International Organization for Standardization (Heiden, telephone conversation, 30 July 1997). In April of 1982 the review board chose Heiden's Defense Data Network plan over AUTODIN II, putting the ARPANET back at the center of the DoD's networking plans (Parker 1982b).

The Transition to TCP/IP

Since the Internet protocols were to serve as the common language for the new Defense Data Network, it became imperative that the ARPANET sites adopt TCP/IP and retire the older and more widely used NCP. After the Internet protocols had been successfully tested on the ARPANET, they would be introduced on the other participating defense networks. In March of 1981, Major Joseph Haughney announced that all ARPANET hosts would be required to implement TCP/IP in place of NCP by January of 1983 (Haughney 1980a, 1980c, 1981a). His successor, Major Glynn Parker, commented on this decision: "Just as it did a decade ago, the ARPANET community is leading the way into a new networking territory of great importance to the future of US military command and control systems." (Parker and Cerf 1982)

The reality beneath Parker's inspiring words was that the DCA and ARPA were forcing a traumatic upheaval in the ARPANET community. Most host sites were still relying on NCP, and converting to the new Internet protocols proved to be an enormous effort. "The transition from NCP to TCP was done in a great rush," one participant recalled, "occupying virtually everyone's time 100% in the year 1982. *Nobody* was ready. . . . It was a major painful ordeal." (Crispin 1991) Dan Lynch, a computer systems manager, recalled: "Dozens of us system managers found ourselves on a New Year's Eve trying to pull off this massive cutover. We had been working on it for over a year. There were hundreds of programs at hundreds of sites that had to be developed and debugged." (Lynch 1991) Lynch made up buttons that read

"I Survived the TCP Transition" and passed them out to his colleagues (ibid.). Alex McKenzie of BBN agreed that there had been a "mad rush at the end of 1982" to make the deadline (McKenzie 1991b). Clearly the transition to the Internet protocols would not have occurred so quickly—perhaps not at all at many sites—without considerable pressure from the military managers.

Most host system managers had no compelling interest in converting to the Internet protocols, and the transition required a number of steps that would cost the host sites time and money. Haughney warned the ARPANET sites in July of 1980: "Unless you have already begun development of the protocols, you may want to start budgeting for the protocol software development for your host" (Haughney 1980a). The first transition occurred in January of 1981, when the new Internet packet format, with 96-bit rather than 32-bit headers, came into use. Hosts had to make sure that all their network applications produced packets with the new headers; if not, they would be unable to use the ARPANET as of January 1981. The next step would be writing TCP software for each type of host computer—which, as the earlier efforts to implement TCP had shown, was no easy task. Hosts would also have to adopt updated versions of the applications protocols ftp and telnet, a new mail standard called Simple Mail Transfer Protocol (SMTP), and a new addressing scheme for mail (Feinler 1982b).[28] At the same time, the sites had to replace their old IMPs or TIPs with new versions designed by BBN to run the Internet protocols (Haughney 1980a; Parker 1982b).[29]

To support inter-network routing, the Internet needed a name server—a database of host names that, when queried with the name of a host, would supply the host's network address. The name server was created by a large group of ARPANET members and went into service at the Network Information Center at SRI in July of 1982.[30] Since NCP and TCP were incompatible, some sites ran both protocols and acted as translators between TCP and NCP hosts during the transition period (McKenzie 1997). Until 1 January 1983 both protocols would also be accepted by the IMPs, but after that date BBN would set the IMPs to reject packets that used the NCP format.

When the cutoff date arrived, only about half the sites had actually implemented a working version of TCP/IP. IPTO director Robert Kahn recalled:

The biggest problem was just getting people to believe that it was real. . . . We sent messages to everybody, alerting them to the timing and yet one week before we were still getting messages, "Is this really going to happen next week?" or "Let us know if you decide to really go ahead with this." (Kahn 1990)

Those who had not created the necessary software for their computers were unpleasantly surprised when BBN upheld the ARPA-DCA policy and cut them off from the network. In addition, many sites that had tried to convert to TCP discovered errors in their implementations and were forced to revert to NCP (Heiden 1983a). Kahn (1990) recalled that it took a long time for all the sites to get their new TCPs working properly:

Managing it was traumatic for a while. I mean, the phone was ringing off the hook every few minutes. Every day someone new would complain, "I used to be able to do this, and now I can't." Shaking it all down was also a problem. Even the places that thought they were going to convert properly suddenly found that while theirs worked with the three or four places that they thought it would, or had tried it out with, it didn't work with some others.

To keep the network running, host sites with nonfunctional TCPs were temporarily allowed to run NCP while they worked on the problem. Any site that had not converted to TCP/IP by the cutoff date was required to submit a request for an exception, justify its failure to be ready, and set a schedule for converting (Heiden 1982). By March of 1983, when the next deadline arrived, about half of the remaining sites still did not have the new protocols running, and the routine was repeated. By June every host was running TCP/IP. A major milestone in the evolution of the Internet had been passed (Heiden, telephone conversation with author, 30 July 1997).

Steps toward a Civilian Internet
After converting the ARPANET to TCP/IP, the DCA and ARPA took two more steps that would help set the stage for the development of a large-scale civilian Internet.

One step was to segregate the ARPANET's military users and its academic researchers, who had been coexisting somewhat uneasily since the DCA's takeover of the ARPANET in 1975. The DCA and its military users were concerned that the academic sites could not or would not enforce strict access controls. One BBN manager put it this way: "The research people like open access because it promotes the sharing of ideas. . . . But the down side is that somebody can also launch an attack." (Broad 1983, p. 13) The DCA warned in 1982 that the ARPANET was increasingly vulnerable to "intrusion by unauthorized, possibly malicious, users . . . as the availability of inexpen computers and modems have made the network fair game for cou...less computer hobbyists" (Harris et al. 1982, p. 78). To protect the military sites from this perceived threat, Heiden decided to split the ARPANET into two separate networks: a defense research network (still called ARPANET) and an operational military network (MIL-NET). The ARPANET would continue to be used to develop and test new networking technologies, while MILNET sites would be equipped with encryption devices and other security measures to support their military functions (Lukasik 1997). The decision to split the network was announced on 4 October 1982, and the MILNET was officially established on 4 April 1983 (Heiden 1983b). The actual physical separation of the two networks took a bit longer. Each host and IMP had to be assigned to either MILNET or ARPANET, and telephone links had to be rearranged so that only IMPs from the same network would be interconnected. A few hosts were attached to both networks to provide internetwork communications (Heiden 1983b). The new arrangement meant that the ARPANET was once again a research-oriented network dominated by universities. This would make it much easier to imagine transferring the network to civilian control.

The second step to was to commercialize the Internet technology. Heiden was eager to have commercial sources for Internet products. ARPA had already funded various contractors to write TCP implementations, most notably for the Unix operating system. Heiden stepped up this effort at technology transfer, setting up a $20 million fund to finance computer manufacturers to implement TCP/IP on their machines (Heiden, telephone conversation with author, 30 July 1997). All the major computer companies took advantage of this opportunity, and by 1990 TCP/IP was available for virtually every computer on the American market. This gave a tremendous momentum to the spread of the ARPA protocols, helping to ensure that they would become a de facto standard for networking.

Ushering in the Internet Era

In the period 1973–1983, ARPA created a new generation of technologies for packet radio, packet satellite, and internetworking. The ARPANET went through a number of transformations: the entire network community switched to TCP/IP, the military users were split off to their own network, and the ARPANET became part of a larger system—the

Internet—that encompassed a number of military and experimental networks. Owing in large part to ARPA's influence, the field of computer networking underwent a conceptual transformation: it was no longer enough to think about how a set of *computers* could be connected; network builders now also had to consider how different *networks* could interact. The dominant model for internetworking would be the system worked out by Vinton Cerf and Robert Kahn.

In the years since the Internet was transferred to civilian control, its military roots have been downplayed. However, it should not be forgotten that ARPA's new networking techniques were shaped in many ways by military priorities and concerns. Like the original ARPANET project, the radio, satellite, and Internet programs followed a philosophy of promoting heterogeneity and decentralization in network systems that mirrored the US military's diverse and scattered operations. The use of new communications media was meant to make it easier to tailor command and control systems to specific military environments, such as jeeps, ships, or airplanes. The idea that network protocols should be simple and adaptable derived in part from the military's continued concern with survivability. Even civilian developments were shaped by the military: Robert Metcalfe drew on the ARPA-funded Alohanet work in developing Ethernet, and Heidi Heiden funded the commercialization of TCP/IP. Finally, it was the determined efforts of DCA managers to get TCP/IP running throughout the ARPANET that set the stage for the emergence of a worldwide, publicly accessible Internet in the late 1980s.[31]

But military shaping is only part of the story. The Internet approach would not have been so influential had it not served the needs and interests of a diverse networking community. The Department of Defense could require the US armed forces to use the TCP/IP protocols, but it could not force others to adopt them. The ARPA system was not the only option available: by the mid 1970s, both computer manufacturers and telecommunications carriers were beginning to offer their own internetworking systems, which might have served as the basis for a worldwide network system. Cerf and Kahn's collaborative approach to system design helped ensure that TCP/IP would become the technology of choice. And, as Cerf (1990) observed, the Internet's ruggedness made it appealing for civilian as well as military applications:

There were all kinds of challenges for this technology to overcome that were military in nature, that were problems that were caused by very hostile environments. Now as it has turned out, the robustness in the system has been helpful in the civilian sector, too. They may not be as dramatic, but a cable cut through an optical fiber line is just as devastating as nuking some central office somewhere, as far as communications is concerned.

By coordinating defense and civilian interests, the Internet's designers were able to create a system that would appeal to a broad spectrum of potential network builders.

The story of the Internet's origins departs from explanations of technical innovation that center on individual inventors or on the pull of markets. Cerf and Kahn were neither captains of industry nor "two guys tinkering in a garage." The Internet was not built in response to popular demand, real or imagined; its subsequent mass appeal had no part in the decisions made in 1973. Rather, the project reflected the command economy of military procurement, where specialized performance is everything and money is no object, and the research ethos of the university, where experimental interest and technical elegance take precedence over commercial application. This was surely an unlikely context for the creation of what would become a popular and profitable service. Perhaps the key to the Internet's later commercial success was that the project internalized the competitive forces of the market by bringing representatives of diverse interest groups together and allowing them to argue through design issues. Ironically, this unconventional approach produced a system that proved to have more appeal for potential "customers"—people building networks—than did the overtly commercial alternatives that appeared soon after.

5

The Internet in the Arena of International Standards

ARPA's Internet program created a set of protocols that would be widely adopted both within and beyond the Department of Defense. The TCP/IP protocols were developed in the 1970s and became a de facto commercial standard in the 1980s. But during this same period, the increasing popularity of computer networks led international standards bodies to propose formal standards for network protocols—standards that did not include TCP/IP. As data networks passed the experimental stage and began to have commercial and national significance, the question of which standards should be adopted by national and commercial networks touched off a heated debate within the computer and communications professions.

Standards are a political issue because they represent a form of control over technology. Interface standards, for instance, can be empowering to users of a technology. If all manufacturers of a device use the same interface (for example, the touch-tone keypad of a telephone), users need learn how to operate the device only once. Standards also ensure that components from different manufacturers will work together. When standard interfaces make products interchangeable, consumers can choose products on the basis of price or performance, rather than just compatibility. This increases consumers' power in the marketplace relative to producers.

Manufacturers have their own interests, which may be opposed to those of users. Large firms such as IBM have often tried to protect their established markets by keeping their internal product standards secret, thus making it difficult for other vendors to offer products compatible with their own (Brock 1975; Lamond 1985).[1] Conversely, smaller manufacturers have an incentive to work for common standards that will remove barriers of incompatibility and put vendors on a more level playing field. Since most national governments actively promote the competitiveness of their domestic industries, technical standards that affect markets can also become matters of foreign policy.[2]

All these factors came into play in the debate over network standards. Standardization had obvious benefits, but the choice of any particular protocol as an international standard would also create winners and losers among the creators and users of network technology. In the 1970s the computer manufacturers controlled the market for network products: there were no commercially available non-proprietary network systems, and the ARPA protocols were still confined to a restricted-access research network that did not yet seem relevant to most computer users in the private sector. Computer users who were unsatisfied with what the manufacturers chose to provide had an incentive to work for public standards. There was also an element of international competition, since US computer makers dominated the world market. Computer users, computer manufacturers, telecommunications carriers, and government agencies all sought to defend their interests by arguing for or against particular types of network standards.

Tracing the history of the standards debate highlights the roles of the various interest groups that were involved in data communications during the formative years of the Internet. The outcome of this standards debate would shape the world of international networking in the 1980s and help set the stage for the Internet's worldwide expansion in the 1990s.

Standards Makers: A Web of Interests

Technical standards come from a variety of sources. Sometimes the first version of a technology to succeed in the marketplace sets the standard, as happened with the QWERTY typewriter keyboard. Standards (such as QWERTY) that have no legal sanction but have become widely used are called "de facto standards." One of the strongest de facto standards for computer networks was ARPA's TCP/IP protocol suite. When efforts to develop international network standards began, in the mid 1970s, TCP/IP was in its infancy; however, the debates continued into the mid 1980s, by which time the TCP/IP protocols were in wide use in government and university networks in the United States, had been adopted by some research networks in Europe, and were commercially available from many vendors. Even before TCP/IP was widespread, the ARPANET's general approach to networking was being championed by US government representatives and by the International Network Working Group, a coalition of network researchers that had participated in the design of the Internet.

Another source of de facto standards was computer manufacturers. In the 1960s and the early 1970s, manufacturers had offered various partial solutions for connecting their machines; however, these had not been systematic—IBM alone had "several hundred communication products," with dozens of different protocols (Tanenbaum 1989, p. 23). In the mid 1970s—perhaps in response to the 1972 ARPANET demonstration, which showed the advantages of a large unified system—several of the major computer manufacturers began to offer comprehensive network systems. IBM came out with its Systems Network Architecture in 1974. In 1975, Xerox introduced Xerox Network Services, and the Digital Equipment Corporation brought out its Digital Network Architecture (DECNET) system. Other computer manufacturers, including Honeywell, Sperry, and Burroughs, followed suit (Bell 1988, p. 16; Passmore 1985, p. 948). All these systems used packet switching and therefore owed a debt to the ARPANET.[3] Unlike the ARPANET, however, these systems were proprietary: they were designed to work only with the manufacturer's own line of computers, the technical specifications were often kept secret, and a license fee was charged for use of the protocols.

Proprietary standards tend to favor large manufacturers and may do little to increase compatibility between different products. To get around these drawbacks, users or producers of a new technology often try to establish formal or public standards. Formal standards are issued by authorized national or international organizations. Unlike most standards that emerge from the private sector, public standards are created with the participation of users as well as producers, and no single company has technical or economic control. Public standards bodies often wait until a technology has been in use for some time and the pros and cons of different designs have become apparent before choosing a standard. However, in other cases—especially where having multiple competing standards would create incompatibility or large conversion costs—public standards organizations try to develop standards before a technology has become firmly established. The latter was the case with the two public network standards that will be considered here.

For computer networks, the public standards situation in the 1970s was quite complex. A byzantine hierarchy of national and international standards bodies had arisen in the past hundred years, and several of these groups claimed jurisdiction over data communications standards. In the United States, the lead in developing formal computing standards was taken by the American National Standards Institute (ANSI), a nongovernment, nonprofit organization that coordinates the development of voluntary standards. ANSI adopts standards by consensus among its members, who represent both users and producers of computer technologies (Henriques 1975, p. 107; Wheeler 1975, p. 105).[4] The National Bureau of Standards,[5] which set standards for US government use, also played a role. Though mainly responsible for standards used within the United States, the NBS sometimes participated in international standards efforts when it perceived that US government users had a strong stake in a particular technology (Blanc and Heafner 1980). Finally, the Department of Defense had its own military standards, which defense contractors were required to follow. All these groups favored US interests when deciding on technical standards.

Internationally, two organizations shared authority for networking standards. One was the Consultative Committee on International Telegraphy and Telephony (CCITT) of the International Telecommunications Union, a treaty organization that had been formed in 1865 to coordinate communications policy among European nations and which later had been authorized by the United Nations to research and develop worldwide technical standards for telegraphy and telephony.[6] The CCITT holds a plenary meeting every four years at which members vote on proposed standards and identify issues for further study. CCITT standards are officially labeled "Recommendations," and in theory they are not binding; in practice, however, they are automatically adopted as national standards by many member countries. Most members are represented in the CCITT by their national telecommunications carriers, which are state-owned Post, Telegraph, and Telephone administrations. Since the United States has no state-owned carrier, it is represented officially by the Department of State, which consults experts from American telephone companies, equipment manufacturers, user groups, and government agencies (Wheeler 1975, p. 105; Schutz and Clark 1974). Private telephone companies and industrial and scientific organizations may participate in CCITT working groups but may not vote. Since the CCITT has always been dominated by telephone carriers, its approach to data networking has been based on its members' experience in telephony, rather than on expertise in computing; this sometimes puts the CCITT at odds with computing advocates.

The CCITT shared authority over networking standards with the International Organization for Standardization (widely known as "ISO"), a body founded in 1946 to coordinate international standards for a wide range of industries. The United States is represented in ISO by ANSI. ISO standards are drafted by Working Groups. Although open to interested representatives from government, from academia, and from user groups, the Working Groups tend to be dominated by manufacturers, who have both a strong interest in shaping the standards for their own industries and sufficient financial resources to send expert employees to participate in standards activities. ISO circulates its Draft Standards among all its member countries for comment and revision; when there is general approval, a draft is officially designated an International Standard. Like CCITT standards, ISO standards are automatically adopted by many nations.

The complexity of the public standards system for computer networks, with numerous overlapping bodies in the United States and internationally, stemmed in part from the traditional jurisdictional divisions between nations and between the hitherto separate fields of computers and telecommunications. But it also reproduced the complex web of interest groups who had stakes in the emerging technology. Rival groups—Europeans and Americans, for instance, or computer manufacturers and telecommunications carriers—turned to their respective preferred standards organizations for the creation of specifications that they hoped would control the future development of networking. The result was a Babel of competing and incompatible "standards."

The difficulty of reconciling these competing interests was exacerbated by time pressures. Standardization proceeded in haste in the 1970s because a large number of public and private groups were beginning to plan network projects. If public standards were not ready in time to be used in these networks, each project might adopt its own local standard, and the result would be chaos. Not wishing to be overtaken by events, the CCITT and ISO created and ratified their initial standards with unaccustomed speed, leaving many technical issues unresolved. Their haste would compound the confusion and contention that surrounded network standards.

This chapter examines in detail the efforts of ISO and the CCITT to create international network standards and the controversy that erupted over these proposed standards. Debates over network standards often proceeded on what appeared to be a purely technical level. Closer examination, however, reveals the economic, political, and cultural issues that underlay these arguments. It was these issues, I argue, that made the standards debate so heated. The standards controversy also illustrates how the field of networking was changing in the late 1970s and the 1980s. The Internet and its creators were no longer operating in the insulated world of defense research; they had entered the arena of commerce and international politics, and supporters of the Internet technology would have to adapt to this new reality.

Recommendation X.25

The first battle over data network standards flared up between the telecommunications carriers and the computer manufacturers. The main bone of contention was whether the market for network products would be controlled by sellers (computer companies) or by buyers (telephone carriers). Setting their own standards for networks would be one way for the carriers to tilt control of the market in their direction. But computer researchers were also drawn into this dispute—especially supporters of the Internet model of networking. The Internet community felt that the carriers were attempting to use their new standards to impose their vision of a worldwide network system on computer owners and network operators, and this vision clashed with the Internet model of networking in several important respects. It was the research community that engaged in the most extensive critique of the carriers' standards in the technical and trade press.

The carriers saw data communications as simply an extension of telephony. In terms of infrastructure, the new public data networks they envisioned would use the existing telephone system, with some additional computing components. The carriers assumed that most customers would use the network to access a computer from a terminal—which, like a telephone, is a fairly simple device. They did not expect most customers to engage in computer-to-computer interactions over the public data networks. As things turned out, the "telephone model" of computer networking did not fit well with the way computer users actually wanted to use networks.

For much of their existence, telephone systems in most countries have been state-owned or regulated monopolies. As computer use increased in the 1960s and the early 1970s, the PTTs in the industrialized world became aware that there was a growing market for data transmission, and they set out to incorporate that market into their existing national monopolies. In 1974 and 1975 the telephone carriers in Europe, Canada, and Japan announced plans to introduce public data networks, which were to be the computers users' equivalent of the public telephone network. Though these data networks would serve only a few large cities at first, they had the potential to grow as large as the telephone networks, and the carriers anticipated that they would eventually interconnect their systems to handle international data traffic.

PTT administrators realized that if they did not explicitly agree on networking techniques they would run the risk of creating incompatible systems that could not easily be interconnected. As early as 1973 the CCITT had begun studying the question of standards for public data networks, and this effort accelerated as it became clear that public data networks would soon be built with or without CCITT standards. The carriers preferred to create their own standards because they did not want to base their networks on computer manufacturers' proprietary products, which might lock them into buying network equipment from a single supplier. They were particularly concerned that IBM would use its huge share of the world computer market to make its new Systems Network Architecture (SNA) a de facto standard.

Tensions between the PTTs and IBM erupted in October of 1974, when the Canadian PTT took a public stand against what it saw as IBM's monopolistic practices. The Trans-Canada Telephone System (TCTS) had begun work on a packet switching network called Datapac, for which it was developing its own protocols. Shortly after this effort started, IBM's SNA protocols became available; since the Datapac network used mostly IBM computers, adopting SNA was an obvious option for TCTS. Instead, the Canadian government issued a statement that it was seeking publicly specified network protocols that would be compatible with a variety of computer equipment. This was widely interpreted in the world of data communications as a direct attack on IBM's policy of keeping its protocols secret and incompatible with rival products. While TCTS tried to persuade IBM to modify its protocols to meet Canada's requirements, IBM urged the carrier to accept SNA, arguing that it did not make sense for a huge corporation to tailor its products for such a small segment of the world market. By mid 1975 the two sides were at a standoff (Hirsch 1976a, 1975).

In response to the Canadian situation and the rapid pace of public network development elsewhere, an ad hoc group within the CCITT decided to create its own protocols for connecting customers to public data networks. This group was led by representatives of the PTTs in Canada, France, and Britain (all of which were developing data networks) and from Telenet (an American commercial data network that was a spinoff of the ARPANET). Starting in 1975, the group began work on a set of three protocols, which they designated Recommendation X.25 ("X" being the CCITT code for data communications standards). The developers of X.25 were forced to work in haste by the timing of the CCITT's standards approval process: proposed standards had to be voted on by the entire CCITT membership, which met only every four years at the CCITT's plenary. If X.25 was not ready in time for the September 1976 plenary, the standard could not be approved until 1980, by which time many of the networks would already have been built. Thus, the X.25 group hurried to produce a first draft in the spring of 1976, and at the September plenary X.25 was approved by a majority of CCITT members and became an international standard.

The PTTs quickly incorporated the new protocols into their developing data networks. Telenet adopted X.25 in 1976, and in 1977 Datapac followed suit. Early in 1978 these two systems were linked, thus demonstrating that X.25 could be used for internetworking. Among the other public data networks to adopt X.25 were France's Transpac (1978), Japan's DDX (1979), the multinational Euronet (1979), and the British Post Office's PSS (1980). With their united support for the new standard, the PTTs were able to pressure computer suppliers to provide X.25 hardware and software to build these new public data networks (MacDonald 1978, p. 258; Davies and Bates 1982, p. 20).

Some manufacturers welcomed X.25, because it created a large and homogeneous market for data communications products. Even manufacturers who had their own proprietary network products hedged their bets by pledging to support X.25 in addition to their own protocols. In 1976, IBM, Digital, Honeywell, and other US computer firms announced that they planned to offer X.25 software for their machines—though, having made this gesture, they did not actually bring X.25 products to the market until years later, when customer demand for the protocols had become too strong to be ignored (Passmore 1985).

Having won their battle with the manufacturers, the carriers might have settled into using X.25 without further incident had it not been for one thing: the Internet, with its own painstakingly crafted protocols and its large community of committed supporters. Many of these supporters felt that X.25 had been set up in direct opposition to the Internet's TCP/IP protocols. One US standards official claimed that the hasty approval of X.25 had been accomplished only with "considerable political pressure through a number of participating nations" (Folts 1978). American representatives to the CCITT had actually suggested adopting TCP/IP as a standard for public data networks, but the idea had been "flatly rejected" (Vinton Cerf, email to author, 17 January 1992).[7] TCP/IP offered tested, freely available, nonproprietary protocols; but the public carriers wanted protocols designed for their own specific needs and interests, which were not necessarily the same as those of the Internet's military and academic computer users. The CCITT created its own standard, and the CCITT and Internet protocols developed along separate paths.

Technically, X.25 and TCP/IP are not mutually exclusive; the two sets of protocols can be and have been combined in a single network. But the ARPA and CCITT protocols had not been designed to work together, and combining them would needlessly duplicate many functions; they were clearly meant to be alternative approaches to building networks.[8] For those watching the development of network standards, X.25 and TCP/IP became symbols of the carriers' and the Internet community's opposing approaches to networking. Though each provided a system for networking computers, they embodied different assumptions about the technical, economic, and social environment for networking. The tension between these two visions manifested itself as a battle over standards.

From X.25's approval in 1976 until the late 1980s, networking professionals carried on an intense debate over the relative merits of the two standards. The stakes were very real: each side wanted its protocols to be supported by manufacturers, taught to computer science students, and chosen for use in public and private networks. The controversy was recorded in a steady stream of articles arguing the superiority of X.25 or TCP/IP in the trade press, in major computing journals (including *Computer Networks, Proceedings of the Institute of Electrical and Electronic Engineers,* and *Communications of the Association for Computing Machinery*), and at international computing conferences. A closer examination of this debate reveals how the technical decisions embodied in standards reflected the beliefs, values, and agendas of the people who designed and used networks.

Virtual Circuits and the Distribution of Control

One of the basic decisions to be made in network design is which part of the system will control the quality of service: the communications subnet, or the host computers (figure 5.1). If a single entity owns the subnet and the hosts, this decision might be made on mainly technical grounds. But if different groups own the different parts of the network, it also becomes an issue of which group—network operators or computer owners—will have the power to determine network performance. Such was the case in the dispute over X.25. To the PTTs, it seemed self-evident that the best service to customers would result from a system where a few large telecommunications providers controlled network operations. To computer owners, it seemed just as obvious that individual computing sites should have the maximum possible control over network performance, so that they could tailor the service to meet their own needs.

The debate over who should control the quality of service focused on an aspect of networking called "virtual circuits." Data travels through a packet switching network as isolated packets that may be lost, damaged, or disordered before reaching their destination. Somehow the network must convert these individual datagrams into a continuous, orderly, error-free stream of data (known as a virtual circuit because it mimics a dedicated circuit between two host computers). Network designers debated whether virtual circuits should be provided by the subnet or by the host protocols (also known as "end-to-end" protocols, since the host computers are the endpoints of the connection). A network that creates virtual circuits within the subnet of packet switches is said to be "connection-oriented," one that simply transmits datagrams to be "connectionless." The carriers and the Internet community came down on different sides of this debate.

The design of X.25 makes the network of switching nodes responsible for providing virtual circuits. This means that the nodes have to set up a connection between two hosts, regulate the flow of data along this virtual circuit, and recover from hardware failures in the hosts or in the subnet (Rybczynski et al. 1976; Dhas and Konangi 1986). This design assumes that the subnet is reliable, and that it includes switching computers capable of keeping track of virtual circuits, controlling congestion, and detecting errors. In the early ARPANET, the subnet

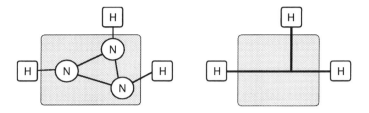

A. Network of switching nodes   B. Ethernet cable or radio link

Figure 5.1
Control over network performance can be vested in the communications subnet (shaded area) or in the host computers. In the ARPANET or the PTT networks, the subnet consists of a set of switching nodes connected by communications links (A). Other types of networks, such as Ethernet LANs and some radio networks, have no switching computers; the subnet is simply a communications link (B).

provided the kind of reliable, error-free service that the CCITT specified. In the later Internet system (with TCP/IP), the subnet was required only to provide datagrams. The host protocol, TCP, was responsible for setting up the connections between hosts and for providing flow control and error correction. The Internet system put most of the responsibility on the hosts and made minimal demands on the network.

These two approaches reflect different assumptions about the capabilities of the network and about the system operator's control over those capabilities. For the telecommunications experts at the CCITT, it seemed reasonable to assume that public data networks—like the telephone networks after which they were modeled—would transmit information reliably. Since all the public data networks were designed with switching computers, it was also reasonable to assume that the subnet could handle the complex operations required to provide virtual circuits. On the other hand, the CCITT could make no assumptions about the host computers, which belonged to the PTTs' customers. Therefore, it made sense for the carriers to concentrate control over network operations within the network. In contrast, ARPA's network designers assumed that some networks might not be reliable: perhaps they had to operate in a hostile environment, or perhaps they used a technology (e.g., packet radio) that was inherently unreliable but offered offsetting advantages, such as simplicity, low cost, or mobility. Since the networks could not always guarantee error-free service, it was safer to rely on the host protocol for virtual circuits. Unlike the CCITT, ARPA could be confident that the hosts on the Internet would use such a protocol, since it had made the use of TCP mandatory for its host sites and had encouraged manufacturers to offer commercial TCP products.

Why should it matter whether the work of providing virtual circuits is done in the network or in the hosts? For some experts, it was a matter of network performance. Critics of X.25 argued that the only way to guarantee reliable network service was to have the end-to-end protocol do the error checking. X.25 depended on each network in an internet to provide high-quality service; if any of them failed to deliver a packet, the host had no way to recover from the error. But if the hosts themselves were responsible for error checking, it would not matter how unreliable any of the networks were: as long as the internet to which they were attached was functioning at all, the hosts could ensure that their messages were delivered intact. The designers of the French Cyclades network, which relied on the hosts for error control, argued: "A major deficiency, especially for the level of complexity of X.25, is the lack of mechanism assuring end-to-end integrity of data flows. This will likely prevent X.25 from becoming the workhorse of data transmission, except where an end-to-end protocol is also provided." (Pouzin and Zimmermann 1978, p. 1367) ARPA personnel pointed out that even a normally reliable network might suffer damage during a war, so a military network would need to have end-to-end error control by the hosts to ensure reliable service under hostile conditions.

But relying on the hosts for performance guarantees had the drawback that computer owners would have to implement a complex protocol such as TCP on their machines. The PTTs felt this was an unrealistic requirement; they expected that users would want simple host protocols and minimal responsibility for network operations. The builders of the Canadian Datapac network wrote that they had "decided that the virtual circuit concept would be adopted . . . to minimize the impact of network services on user systems" (Twyver and Rybczynski 1976, p. 143). In view of how difficult it turned out to be for the ARPANET sites to implement TCP, it was probably reasonable for the PTTs to avoid making such demands on their customers. On the other hand, once ARPANET users had invested the effort to implement TCP it made sense to take full advantage of that protocol's capabilities. Thus, the appeal of having the hosts provide virtual circuits depended, at least in part, on whether the protocols in question were already available or whether they would have to be implemented (perhaps with great effort) by the system's users.

The more technically sophisticated users were often willing to take on this additional responsibility in return for greater flexibility and control over the system. Academic and commercial researchers and managers of large computer installations tended to want protocols that would put control of network performance in their own hands. During the X.25 design meetings in the spring of 1976, *Datamation,* a magazine for data processing managers, reported that end-to-end protocols had become the focus of a power struggle between computer users and PTTs:

There is a heated international argument over who will control packet switched communication networks—the carriers or the users. . . . Many multi-terminal users believe they can maximize the benefit of packet service only by employing end-to-end communication protocols designed for their specific systems. This contention makes the carriers livid and helps explain why the argument was gathering heat at the Geneva [CCITT] meetings. (Hirsch 1976a)

Data processing managers wanted to be able to tailor the network's behavior to their organization's needs, which meant having their own hosts do more of the work. These expert users tended to regard virtual circuits in the network as an expensive redundancy. The trade journal *Electronics and Power* reported:

Needs of computer users vary so much that there is difficulty in fixing . . . a technical specification that will be attractive to all. . . . Some users will want error control done for them; others, who are in a position to provide end-to-end error control of their own, will be less desirous of paying for error control also to be done in the network, nor will they wish to accept the transmission delays to which it may give rise. (Wilkes 1980, p. 70)

Transmission delays would be especially burdensome in real-time applications, such as packet voice or video.

To avoid unnecessary expense and delay, a number of computer users wanted the option of forgoing virtual circuits altogether in favor of unadorned datagram service, in which packets would be transmitted independently with no attempt to maintain an orderly sequence of data. In 1977, ANSI issued a "USA Position on Datagram Service" arguing that for certain applications datagrams would be preferable to virtual circuits. These included applications where speed was critical (so users would rather not wait for the network to perform error checks), applications where messages were short enough to fit into a single packet (so that the overhead of setting up a virtual circuit would be wasted), and situations where the network design had to be very simple. After "an intensive effort by ANSI" (Folts 1978, p. 251), the CCITT added an optional datagram interface to the X.25 specification. However, private network builders needed special permission from their carrier to use this option, and the PTTs proved unwilling to provide anything but the more expensive virtual circuit service (Tanenbaum 1989, p. 322; Hirsch 1976a; Pouzin 1975a). Many private network owners suspected that the PTTs saw private networks as competition and had no intention of aiding them. *Datamation* predicted that in the United States competition between carriers would eventually force them to provide the datagram option to users who wanted it, but that "in Europe, where government policy favors operation of a single network within each country, the situation is considerably different" (Hirsch 1976b, p. 190). "By refusing to allow use of some or all datagram protocols instead of X25," *Datamation* continued, "the PTTs—which are largely or completely government-owned—conceivably could limit the development of private networks." As it turned out, none of the public carriers were willing to implement the datagram option, and in 1984 it was removed from the standard.

In defending their decision to provide only virtual circuits, the carriers argued that computer owners and private network builders did not understand the reality of operating a commercial network service. Though private network owners could choose to build unreliable systems, the public had always demanded reliability from the phone system and would expect public data networks to be dependable. Therefore, the PTTs felt it necessary to provide virtual circuits rather than just datagrams. They also wanted to concentrate functionality in the part of the network that they controlled. The designers of the French public data network, Transpac, commented: "Mechanism for transporting datagrams appears in the architecture of various *private* networks. However, for a public network, new problems appear due to the separation of responsibilities at the interface with subscribers." (Danet et al. 1976, p. 252) The CCITT's protocols were deliberately designed to put control of the network in the hands of the PTTs by locating most of the functionality within the network rather than in the subscribers' host computers. The PTTs asserted that the public would hold them accountable for the performance of the system.

The operators of public data networks argued that ARPA's TCP/IP failed to provide adequate control over network operations. For instance, a Telenet spokesman noted that, whereas X.25 was capable of controlling the flow of packets from each individual connection, TCP could only act on an entire host's output at once. If one of the network connections from a host malfunctioned and flooded a TCP/IP network with packets, the network's only defense would be to cut off the entire host, thus unfairly penalizing the other users on that host (Roberts 1978, p. 1310). Users of a research network might accept the inconvenience with resignation, but paying customers of a public data network would certainly protest. With regard to the business of running a network, the PTTs pointed out that IP had not been designed to allow networks to exchange the type of information that would be required for access control or cost accounting (Landweber and Solomon 1982, p. 401). Charging users for network services had never been a priority for ARPA. One ARPA contractor, Franklin Kuo, explained in 1975: "During the early days of the ARPANET, ARPA paid the entire communications and computation bill. . . . At the time of this writing, no network-wide accounting plan has yet been instituted." (Kuo 1975, pp. 3–15) TCP/IP had not been designed for a network serving as a public utility, with service guarantees and access charges; X.25 had been.

The choice between X.25's virtual circuits and TCP/IP's datagrams was not simply a technical matter; it also shifted the distribution of control and accountability between public network providers and private computer owners. For the PTTs, virtual circuits meant they could guarantee their customers better service and boost their own profits. Some of their customers were glad to be offered reliable data communications service with little effort required on their part. For more expert computer owners, however, virtual circuits raised the cost of network service and interfered with their ability to control their own data communications activities.

Internetworking and the Role of Private Networks
A second and related topic of debate involved internetworking. Each of the two systems provided a way to interconnect networks, but these methods presupposed very different models of the resulting system. In the ARPA model, the interconnected networks would remain distinct and would retain the option of using different protocols to transmit packets internally. Gateways between networks would forward

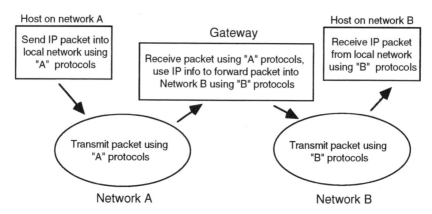

Figure 5.2
The ARPA internetworking scheme.

packets from one network to another, using each network's local protocols, and IP would provide the common packet format and routing
procedures (figure 5.2). In the PTT model, the set of connected
networks would behave like a single homogeneous network. In fact,
there was very little distinction in the CCITT model between connecting a single host to the system and connecting an entire network. Since
X.25 provided a common format for all the networks, the carriers
thought no additional internet protocol would be necessary. The
CCITT did eventually add an internet protocol (called X.75), but this
did not substantially alter its original model. X.75 was only a slight
variation of X.25 and was only intended to work with X.25 networks;
it was not meant to be a way to link diverse networks. Gateways were
optional in the CCITT system: an X.75 gateway could be set up
between networks, or a link could be made between the two networks
using an ordinary X.25 node (figure 5.3).

It is evident that the Internet designers expected to accommodate
a diverse set of networks, while the carriers expected every network
in their system to use X.25. The PTTs' model was the telephone
system, and they assumed that their monopoly on telecommunications
would allow them to create a single, homogeneous public data network
in each country. Building a uniform worldwide data communications
system would have obvious advantages for the carriers—and, arguably,
for their suppliers and customers. It would make it much easier to
interconnect the various national networks, and it would make it
possible to create standard network interface equipment that customers would be able to use with any network (Rybczynski et al. 1976,

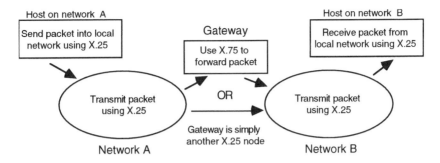

Figure 5.3
The CCITT internetworking scheme.

pp. 7–8; Danet et al. 1976, p. 253). The PTTs also expected that public and private networks would eventually adopt the same international standards, thereby eliminating any need to internetwork diverse systems. Regarding network diversity as an obstacle to creating an efficient internet system, they did not see any compensating advantages to having different types of networks. As late as 1983 a researcher working for the National Bureau of Standards observed "a difference of opinion [among PTTs] as to whether the existing diversity of networks is a temporary unfortunate circumstance, or an inevitable permanent condition" (Callon 1983).

In contrast, ARPA not only expected but welcomed diversity in networking. Just as the ARPANET had connected many different kinds of computers, so ARPA's Internet Program was aimed at connecting many different types of networks. In part this reflected the changing telecommunications environment in the United States during the 1970s and the 1980s. Deregulation was opening up AT&T's telephone monopoly to competition; unlike the state-owned carriers in Europe, Japan, and Canada, AT&T could no longer expect to set standards unilaterally. But ARPA's attitude was not just a matter of adapting to the "unfortunate circumstance" of diversity; ARPA managers viewed the ability to have a variety of networks as a positive good, because it allowed networks to be specialized for particular military environments. In addition, there were an increasing number of academic, government, and commercial networks in the United States that ARPA wanted to be able to include in the Internet.

To accommodate diversity, the ARPA protocols had been designed to make few demands on the subnet; the aim was that any kind of network, no matter how limited its capabilities, should be able join a

TCP/IP-based internet. The X.25 system, which was designed with the expectation that every network would provide a high level of service, put limits on the types of networks that could be accommodated in an internet. These limits became glaringly apparent with the popularity of local-area networks in the early 1980s. LAN systems such as Ethernet, token ring, and token bus do not have computerized switching nodes; instead, they broadcast messages across a medium (such as a cable or a radio band) that is shared by all the hosts. Broadcasting reduces the cost and complexity of LAN technology: no routing is needed, since every host receives every message. But without computerized switches, there is also no capability for providing advanced services within the network, so it is difficult to make a LAN provide the virtual circuits required by X.25 (Quarterman 1990, p. 418; Carpenter et al. 1987, p. 86). The same was true of other broadcast networks, such as ARPA's experimental packet radio network. A 1983 report describing military requirements for networks pointed out the mismatch between X.25 and broadcast networks: "While it is possible to interoperate with these broadcast media using a circuit-like protocol, it is awkward and inefficient to do so. Thus, exclusive use of virtual-circuit protocols fails to utilize inherent capabilities of these broadcast media which have been acquired at considerable effort." (Cerf and Lyons 1983, p. 298) The ARPA Internet community, which included many sites with Ethernet LANs, believed that an internet should be able to include broadcast networks; the PTTs, in contrast, ignored or rejected the possibility that private LANs would be connected to the public data networks.

The differences between the X.25 and TCP/IP internetworking approaches went beyond the issue of network diversity. The design of the CCITT internetworking scheme suggests that the carriers were reluctant to interconnect their systems with private networks at all, even if the private networks agreed to use X.25. The carriers created a double standard for public and private X.25 networks with respect to internetworking. The CCITT allowed the X.75 internetwork protocol to be used only for connections between two public networks; private networks had to connect to the public networks using X.25. What did this mean for private network operators? The chief difference between X.25 and X.75 is that an X.25 host or gateway always represents the termination of a virtual circuit, whereas an X.75 gateway can be placed in the middle of a virtual circuit (figure 5.4). This means that when two networks are linked by an X.25 gateway, an

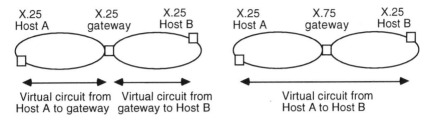

Figure 5.4
Internetworking with X.25 (left) and with X.75. Use of X.25 at the gateway creates two virtual circuits; use of X.75 at the gateway creates only one.

internet connection between two hosts must use two virtual circuits: one from the first host to the X.25 gateway and one from the gateway to the second host. When the networks are connected with X.75, there is only one virtual circuit from source to destination. With fewer virtual circuits, transmission is more efficient and more reliable; there is less overhead and less chance for circuit failure (IFIP Working Group 6.1 1979, p. 37). In practice the difference in performance might be small, but the CCITT's restrictions on the use of X.75 reinforced the message that private networks were not full and equal members of its internet system.

The PTTs did not see any need to interconnect their systems with large numbers of private networks. They envisioned that each country would have a single public data network, and that the various public networks would interconnect at national borders. Computer owners would attach their machines directly to the public data networks, rather than to private networks linked to the public system. In 1978 the British PTT offered this prediction: "For many types of data communications public data networks are likely to offer a more reliable and lower cost data service to the public than using the international telephone network or [building private networks from] leased lines." (Kelly 1978, p. 1548) In designing the X.25 standard, the PTTs' technical and policy decisions were influenced by this expectation (or hope) that the use of private networks would not be widespread.

One serious issue that arose from the CCITT approach concerned network addressing. Each host within a network has a unique address, analogous to a telephone number; packets sent to that host carry its address in the packet header, which the switches use to route the packet to its destination. Within an internet, each network must also have a unique address, much as long-distance telephone calls require an area code as well as a phone number. But the CCITT addressing scheme allocated very few network addresses for private networks, because the carriers assumed that most users would rely on public networks. The potential shortage of network identifiers was particularly evident in the United States, where dozens of private local and regional networks were being built. LAN technologies such as Ethernet were already under development by the late 1970s, and these would soon make it practical for every university or business to have its own network. Yet the CCITT decided that most countries would require only ten network addresses. The United States, with its abundance of networks, was allotted only 200. Vinton Cerf of ARPA observed: "It might be fair to assume that the United States will not need more than 200 *public* network identifiers. However, this scheme does not take into account the need for addressing private networks." (Cerf and Kirstein 1978, p. 1400) Andrew Tanenbaum, the author of a widely used textbook on computer networking, contended that, because the CCITT represented carriers rather than computer users, it had overlooked the needs of private networks. He argued that the address-shortage problem was "not one of poor estimation" but "a question of mentality." "In the CCITT's view," Tanenbaum continued, "each country *ought* to have just one or two public networks. . . . All the private networks do not count for very much in CCITT's vision." (Tanenbaum 1989, p. 322) In contrast, even ARPA's earliest Internet system, proposed in 1974, had been designed to address up to 256 networks (Cerf and Kahn 1974, p. 99). By the early 1980s, when the Internet still connected only about two dozen networks, ARPA's technical planners were already assuming that a thousand or more networks might soon be included, and they decided to make room for almost indefinite growth in the Internet's address space (Clark 1982). The Internet group revised the IP address system to provide identifiers for more than 16,000 "large" networks (those with hundreds or thousands of hosts each) and more than 2 million "small" networks (128 or fewer hosts). While the unexpectedly rapid growth of the Internet put pressure even on this expanded addressing system by the mid 1980s, the principle remained that all private networks should be accommodated.

Lessons of X.25
Beneath the seemingly dry and technical details of the X.25 standard lay some overt and hidden economic and political agendas. X.25 was

explicitly designed to alter the balance of power between telecommunications carriers and computer manufacturers, and in this it succeeded. Public data networks did not have to depend on proprietary network systems from IBM or any other company. Instead, by banding together, the carriers forced the manufacturers to provide network products based on the CCITT's standard. The X.25 standard also, intentionally or not, pitted computer experts against telecommunications professionals, and private network operators against public carriers. Some of these tensions may have arisen from a lack of understanding on the part of the CCITT's protocol designers of the needs of computer owners, or from the fact that the protocols were developed in a rush without adequate time for discussion of alternatives (Pouzin 1975a). But the standards debate also revealed conflicting assumptions about how data networks should be used and who should control their operation. The carriers intended to create a centralized, homogeneous internet system in which network operators controlled network performance; they also tried to perpetuate their monopoly on communications by making it difficult for private networks to connect to the public systems. This system design would ensure the carriers a large and profitable market for high-quality data communications services. Computer owners, recognizing that the X.25 system would limit their options for customizing network service to meet their military, research, or business objectives, demanded the freedom to choose the level of the service they would purchase from the public networks and to build their own private networks using a variety of techniques.

The conflict between the CCITT and Internet visions of networking was not immediately resolved. X.25 was adopted for most public and some commercial data networks, while the ARPA Internet and many private networks continued to use TCP/IP or commercial protocols. But even as the question of how to interconnect these various systems was being debated, the X.25 issue was overshadowed by an entirely new development in network standards.

Open Systems Interconnection

The second major standards effort, that of the International Organization for Standardization, also took aim at computer manufacturers. Incompatibility between computer systems meant that computer manufacturers essentially had a monopoly over network products for users of their systems, leaving computer users with a limited selection of network products and a continuing inability to network different types of computers. A few years after the CCITT began to work on X.25, therefore, a group of computer experts within ISO launched a campaign to create a set of network standards that would be usable with any computer system.

The network standards effort was a departure from ISO's usual practice, in that it represented an attempt to standardize a technology that was still new and had not had a chance to stabilize. Unlike the CCITT, which developed standards for a relatively small and homogeneous group of carriers, ISO was trying to serve the entire diverse worldwide community of technology users. ISO normally took a conservative approach, waiting until a de facto standard had emerged from practice in a given field and using that as the basis for a formal standard. But in the case of networks, some ISO members felt that formal standards should be outlined proactively. The controversy between the Canadian PTT and IBM (which had helped spur the development of X.25) seemed to demonstrate that, though users wanted compatible network products, computer manufacturers were not committed to providing them. If network users made no effort to develop worldwide standards, they would be faced with a plethora of competing proprietary "standards" that would apply only within each manufacturer's product line. The leader of the ISO effort, the French network researcher Hubert Zimmermann, warned: "The 'free way' to compatibility is homogeneity: let your beloved manufacturer do it, he will always provide compatible equipment but he will also do his best to prevent any one else being able to do it!" (Zimmermann 1976, p. 373)

Early in 1978, concerned ISO members from the United States, Britain, France, Canada, and Japan formed a new committee to tackle the problem of network standards, naming their project Open Systems Interconnection. To computer users, "open systems" represented an ideal that was defined in opposition to the manufacturers' proprietary systems. To maximize their control over the market, computer manufacturers tended to keep their systems as "closed" as they could: they kept the technical workings of their systems hidden from competitors, used patents and copyrights to prevent others from duplicating their technology, made it hard to interface their equipment with components from third parties, and reserved the right to change their "standards" at will. In an open system, by contrast, the technical specifications of the system would be public. The technology would be non-proprietary, so that anyone was free to duplicate it; the system would be designed to work with generic components, rather than only with a specific manufacturer's products; and changes to the standards would be made by a public standards organization, not a private company (ISO Technical Committee 97, Subcommittee 16 1978, p. 50).

By making computer products more interchangeable, openness shifts a measure of control over the technology from producers to consumers. One corporate data processing executive commented after the emergence of Open Systems Interconnection: "To the extent that open systems produce commodity computing, that's what customers like us want." (Carlyle 1988) ISO's challenge to the computer manufacturers went even further than the CCITT's. X.25 had been aimed at forcing manufacturers to provide non-proprietary protocols, but X.25 had only been mandated for the national data networks. Open Systems Interconnection, on the other hand, would be adopted as a national standard in many countries, and therefore many government and private computer buyers could be expected to insist on having OSI products.

Since the field of computer network was still evolving, the OSI committee did not want to draft specific network standards that might prematurely freeze innovation. Instead, they proposed a general framework for future standards development. Their plan was to start by developing a model of how a set of protocols should fit together to form a complete network system; the OSI model would define what services the protocols should provide and how they should interact. Once this overall framework was in place, interested organizations (such as the CCITT or ANSI) could propose specific standards that fit into the model, and ISO's members would decide whether to adopt these as official Open Systems Interconnection standards. The OSI model would be a "meta-standard"—a standard for creating network standards.

To provide a framework for developing standards, the OSI model organized the functions of a network into seven layers of protocols (figure 5.5). The lower layers handle the more concrete tasks of transmitting electrical signals over a physical medium, while the higher layers deal with more abstract matters of organizing and monitoring flows of information. The first three layers of the OSI model—physical, link, and network—are roughly equivalent to the communications

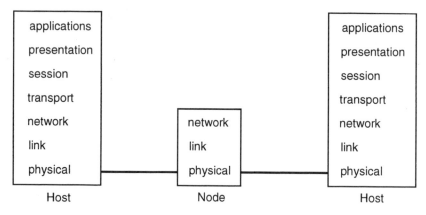

Figure 5.5
The OSI model, showing protocol layers for hosts and nodes.

layer in the ARPANET layering scheme (described in chapter 2). The physical layer specifies how the network interface hardware will regulate the physical and electrical aspects of connections between machines. The link layer translates the flow of electrons across the physical medium into an ordered stream of bits, and decides when to transmit or receive messages from the medium. The network layer handles addressing, routing, and the host-network interface. These first three layers operate on both hosts and switching nodes. The next four layers—transport, session, presentation, and application—apply only to the hosts. The transport layer is equivalent to the ARPANET host layer, providing end-to-end control functions. (TCP is a transport protocol.) The session and presentation layers have no ARPANET counterparts; they provide enhancements over transport service. The application layer is the same as in the ARPANET; it provides specific services, such as file transfer, remote login, or email.

This seven-layer model was supposed to constitute a set of niches into which any proposed ISO protocol would fit. The OSI committee, which saw its work as complementary to other standards movements, stated that its model would allow network builders to "position existing standards in perspective," to "identify areas where standards should be developed," and "expand without disrupting previously defined protocols and interfaces" (ISO Technical Committee 97, Subcommittee 16, 1978, p. 50). The committee hoped that this approach would rationalize the process of choosing network standards without imposing rigid specifications on a field that was still young. In practice, this turned out to be a difficult balance to maintain. When faced with the need to reconcile diverse proposals for network standards, ISO—which relied on consensus among its members to set standards—tended to err on the side of approving too broad a range of standards, rather than restricting its selection of protocols too narrowly.

The Impact of OSI

In proposing that network standards be based on a comprehensive model of the network as a system, the Open Systems Interconnection movement significantly shaped the way computer science professionals thought about networks. The OSI model came to dominate all subsequent attempts to discuss network protocols. Networking textbooks were organized around the OSI layers. Even people discussing protocols that predated OSI, such as X.25 or TCP/IP, dutifully drew charts showing how these protocols fit into the OSI layering scheme. Robert Kahn, one of the designers of TCP/IP, commented that, although the ARPANET had used layering as a practical strategy, the OSI reference model had drawn widespread attention to the idea: "It gave people a way to think about protocol layers. It had certainly been in our consciousness, but we have never articulated it quite that way, and they did." (Kahn 1990, p. 41)[9] OSI enshrined the principle of layering and set up a particular set of layers as the norm. Even manufacturers who continued to use their own protocols adopted ISO's general framework; several manufacturers revised their existing proprietary network systems so that the features of at least the lower layers (physical, link, and network) coincided with the OSI model (Blackshaw and Cunningham 1980, p. 420).

ISO's authority in the field of standards was such that the OSI framework was quickly endorsed by standards bodies in all the countries that were involved in computer networking. In the United States, the National Bureau of Standards formed a Program in Open Systems Interconnection, and in 1983 it began sponsoring workshops to help computer manufacturers implement OSI protocols (Blanc 1986, p. 32). Even the Department of Defense moved toward a policy of using OSI, declaring in 1987 that OSI protocols would eventually be adopted as military standards (Latham 1987). In October of 1981, the ARPANET adopted an OSI link protocol, HDLC, to replace the non-standard link protocol that had been created by Bolt, Beranek and Newman (Perillo 1981).

OSI was particularly welcomed in Western Europe, where the adoption of common technical standards was to contribute to the ongoing effort to integrate the European economies by creating a uniform market for computer products. Public standards were seen as likely to strengthen the European computer companies (which were too small to impose their own proprietary standards) and thus to counter the dominance of IBM and other American manufacturers in the world market. "In Europe," one European involved in standards observed, "independently defined industry standards are seen as the last hope for saving what remains of the indigenous computer industry." (Lamond 1985) In the summer of 1978, shortly after the OSI work began, the French government issued a report recommending that the European nations accelerate efforts to create common data communications standards, arguing: "Control of the network market conditions control of telecommunications and the behavior of the computer market. . . . If IBM became master of the network market, it would have a share—willingly or unwillingly—of the world power structure." (*Electronics* 1978, p. 70)

As OSI's creators had hoped, US manufacturers began adopting the new standards, partly under pressure from users and government authorities and partly to take advantage of the emerging market for open systems. Some smaller manufacturers rewrote their network protocols to conform to the OSI specifications. Even IBM felt pressure from its customers to offer OSI products (such as X.25, which ISO had adopted as an official protocol for the network layer). In 1985, *Datamation* offered this observation: "In the past IBM had hoped SNA would achieve complete dominance in the network architecture arena before the OSI standards organizations got their act together. . . . Due to user demand, however, IBM grudgingly provided an X.25 interface for SNA host systems." (Passmore 1985, p. 105) By offering users the alternative of a publicly defined standard, the OSI effort did, at least in part, achieve its aim of shifting the balance of power between users and vendors.

OSI and the Standards Debate

The 1978 introduction of the OSI model gave a new twist to the ongoing debate over international network standards. ISO took a stance in between the CCITT and ARPA approaches and hence got partial support from both sides. The CCITT and ISO, which had a good deal of overlap in their membership, formally collaborated by sending delegates to each other's committee meetings, and ISO often adopted CCITT recommendations as standards (Wheeler 1975, p. 105). ARPA had no such formal relationship with ISO, but the two groups shared a concern with and an understanding of the needs of computer users.

ISO's approach to network standards was similar to the CCITT's in some respects, especially in the belief that network service should be based on virtual circuits rather than datagrams. ISO was quick to adopt X.25 as an official OSI protocol for the network layer, and this helped gain the carriers' support for the Open Systems effort. But in other ways the two organizations had very different priorities. The CCITT was dominated by carriers, and its members wanted to preserve their centralized control over the national communications networks. ISO's membership, on the other hand, tended toward computer manufacturers and users. ISO promoted a modular, flexible system that would accommodate manufacturers' various approaches to networking and also allow users to "mix and match" network products and services. While the OSI group sided with the carriers on the lower layers of network standards, it was closer in spirit to the Internet users when it came to higher-level standards. The Internet itself could be considered an "open system": its protocols were non-proprietary and freely available, and they were designed to include, rather than exclude, networks of different types. The fact that the Internet protocols had not been established by an official standards body, and the fact that they came from the United States (which already dominated the computing market), made it politically impossible for them to be accepted by ISO as Open Systems standards. But ISO and ARPA shared an understanding of the needs of computer users and a commitment to supporting heterogeneous network systems, and they were willing to work together to bring their standards into closer alignment.

Many in the Internet community took issue with the approach to networking that was implicit in the initial version of the OSI system. The issue of virtual circuits versus datagrams came up again, since ISO had adopted X.25 rather than a connectionless datagram protocol as its first network-layer standard. By the early 1980s, ARPA was also arguing that the OSI model needed an internet layer. In the 1970s, neither the CCITT nor ISO nor ARPA had had a separate internet protocol in its network design; internetworking functions were carried out as an extra duty by one of the other network protocols. In 1980, however, ARPA had split TCP into two separate protocols, TCP and IP, so that the internetworking functions would be separate from the host functions. This had made the Internet independent of any particular network's host protocols, thereby promoting the autonomy and diversity of the member networks. The separate internet layer became a fundamental part of ARPA's approach to internetworking, and Internet users would not accept OSI without it.

One way the Internet community responded to these perceived shortcomings was by becoming more involved in ISO's ongoing efforts to define the OSI standards. In a 1983 survey titled "Military Requirements for Packet-Switched Networks and Their Implications for Protocol Standardization," Vinton Cerf and Robert Lyons argued that TCP/IP users should have worked harder to have their views included in the initial OSI model:

The reason the military finds itself wishing that the ISO model for Open Systems Interconnection incorporated an internet protocol layer, is because they have not convinced ISO and CCITT of this requirement. Similarly, if there is a sufficient need for a connectionless protocol to parallel the CCITT X.25 virtual circuit protocol, then military users and others who share that need ought to be able to convince the international standards setting bodies of that requirement. . . . Perhaps it can be said that military planners did not take the movement toward data communication standards seriously enough. (Cerf and Lyons 1983, p. 304)[10]

US military and civilian Internet users quickly stepped up their involvement in the international standards process. Led by ANSI and the NBS, Americans participated in technical committees and lobbied for acceptance of the Internet protocols as OSI standards.

As a first step toward meeting the needs of Internet users, US representatives in ISO tried to get TCP accepted as an OSI transport-layer protocol. ISO members from other countries rejected this idea, however, apparently fearing it would give an undue advantage to US manufacturers.[11] Instead, ISO decided to define a set of new transport protocols, which were designated (in ascending order of complexity) TP0, TP1, TP2, TP3, and TP4 (ISO 1984a). Working within the system of ISO technical committees, Americans, led by representatives from the National Bureau of Standards, arranged to do much of the design work for the TP4 protocol. Under their direction, TP4 was modeled on TCP and included most of its features (Quarterman 1990, p. 67; Blanc 1986, p. 28). The NBS also made sure that users of local-area networks, who were an important part of the Internet community, would be represented in the ISO standards. Thanks to the NBS, the three main LAN standards—Ethernet, token ring, and token bus—became official OSI standards for the link layer.

Finally, there was the issue of whether there would be an internet protocol. ISO's concern with computer users made that organization more sympathetic than the CCITT had been to the internetworking ideas suggested by ARPA and other TCP/IP users. In 1984, not long after the ARPANET had switched over to TCP/IP, the NBS representatives persuaded ISO to add an internet layer to the OSI model (Blanc 1986, p. 28; Callon 1983, p. 1388). Members of the Internet community worked on the specifications for the ISO internet protocol, which was based on ARPA's IP and which became known as ISO-IP.[12] By the mid 1980s, therefore, Internet users were able to get versions of all their most important protocols sanctioned as international standards.

ARPA also developed a technical strategy for dealing with the proliferation of "standard" protocols from the CCITT, from ISO, and from computer manufacturers: it expanded the role of Internet gateways, whose main task is to route packets from one network to another, to include translating between different network protocols.[13] The Department of Defense and the NBS collaborated on developing gateways to mediate between TCP/IP and OSI networks (Cerf 1980, p. 11), and in the early 1980s ARPA experimentally linked different mail systems with special gateways that decoded and re-encoded mail messages between networks (Postel, Sunshine, and Cohen 1982, p. 978). In general, translation gateways are not a perfect substitute for shared standards: they work well only when the two protocols to be translated offer similar services, and even then the need to reformat data can create a bottleneck at the gateways (Blackshaw and Cunningham 1980, p. 421). However, in view of the slow adoption of international standards, ARPA saw translation gateways as a useful interim measure that furthered its overall aim of building an internet that could adapt to changing circumstances and accommodate diverse networks.

ARPA's translation gateways also neutralized X.25 as a rival networking paradigm. For example, in 1982 ARPA demonstrated an experimental gateway that provided an interface between the TCP/IP-based Internet and the commercial, X.25-based Telenet (Blanc 1986, p. 36). In the combined system, the Telenet hosts ran TCP/IP over the lower-level X.25 protocols. The CCITT had designed X.25 to be the primary provider of end-to-end network control; by running TCP/IP over X.25, ARPA reduced the role of X.25 to providing a data conduit, while TCP took over responsibility for end-to-end control. X.25, which had been intended to provide a complete networking service, would now be merely a subsidiary component of ARPA's own networking scheme. The OSI model reinforced this reinterpretation of X.25's role. Once the concept of a hierarchy of protocols had been accepted, and once TCP, IP, and X.25 had been assigned to different layers in this hierarchy, it became easier to think of them as complementary parts of a single system, and more difficult to view X.25 and the Internet protocols as distinct and competing systems.

Lessons of OSI

The Open Systems Interconnection effort re-framed the debate over network standards and provided some guidance for designing and choosing network protocols. The OSI protocols gained wide acceptance outside the United States. But, despite the good intentions and hard work of many people involved in ISO, the OSI model failed to fulfill its promise of providing universal compatibility. The Internet continued to use TCP/IP, and OSI did not succeed in replacing proprietary network systems. IBM and other manufacturers continued to market their own protocols, offering OSI products only when consumer demand forced them to. Many third-party vendors supported a number of protocol systems, including OSI, TCP/IP, and the more popular proprietary protocols, including DECNET and SNA (McWilliams 1987). As a result, compatibility between networks using different protocols would remain an issue.

Some computer professionals argued that ISO's whole approach of using a comprehensive model to shape the development of standards was overly ambitious. Many network designers felt that the OSI model was too complex—that to have so many layers was unnecessary and inefficient. For instance, the session and presentation layers, which had no counterpart in the Internet or in most commercial networks, were for the most part simply ignored. Another common complaint was that the OSI model was empty: the layers were specified, but for many years no actual protocols had been approved to fill most of those layers. ISO was slow to choose specific protocols for the OSI layers—in some cases because there was disagreement over the choice of protocols, in other cases because some of the layers had not existed in previous networks and so no protocols had ever been devised for them. Users who could not wait for ISO to finish deciding on its standards had no choice but to build networks using other, potentially non-standard protocols.

Even worse than the delay in specifying standards was that ISO ended up sanctioning multiple protocols for some of the layers. When multiple protocols had attained large constituencies of users, ISO tended to approve them all as standards. Almost any protocol could be proposed as an OSI standard as long as its developers were willing to give up proprietary control over it; for instance, local-area network techniques that had been created by Xerox, IBM, and General Motors were all adopted as OSI standards. Adopting several alternative standards eased some of the potential conflict over OSI, but in practice having multiple "standards" for each layer made it possible to use OSI protocols to build quite different—and incompatible—systems. The PTTs, for instance, tended to implement only the protocols that fit their own model of networking, such as X.25 and the most minimal transport protocol, TP0. Internet users, if they used the OSI protocols at all, opted for those that most resembled TCP/IP: TP4 and ISO-IP (Quarterman 1990, p. 433). PTT networks and the Internet may both have been "open systems" in the sense that they used OSI standards, but it did not follow that "interconnection" would come easily.

Thus, despite a flurry of standards efforts, by the early 1980s neither OSI nor any other set of protocols constituted a single worldwide standard. OSI's greatest impact may have been on the way people thought about network design. The model became one of the axioms of networking, so that it began to seem natural and inevitable that networks should be organized into a hierarchy of protocols, each layer performing certain defined functions. To a computer science student who learned about networks by studying this model, alternative ways of constructing a network might seem unthinkable. And by placing individual protocols in a framework that covered all aspects of a network, the OSI model forced protocol designers to be more conscious of how the behavior of each protocol would affect the entire system.

Conclusion

The standards debates forced the Internet's supporters to articulate their own networking philosophy, persuade others of its merits, and fight to have the ARPA approach embodied in international standards. The Internet group had worked out its techniques with little input from the commercial world. The CCITT and ISO efforts made it clear, for the first time, that ARPA's model of internetworking would have to compete with other firmly established—and in many ways incompatible—communications paradigms.

The standards efforts of the 1970s and the 1980s helped shape the computing environment in which the Internet would develop. Unlike proprietary standards, X.25 made it possible to build networks that included different types of computers. In this way, X.25 contributed to the trend of heterogeneity within networks that the ARPANET and Ethernet had started. At the same time, the manufacturers' resistance to X.25 contributed to heterogeneity *among* networks: computer vendors continued to offer their own network products as alternatives to X.25, and this helped thwart the carriers' hopes of establishing a uniform, worldwide X.25-based internet. The Open Systems effort by ISO reinforced both of these trends. "Open" protocols were meant to overcome incompatibilities between computers, and they provided another alternative to proprietary network systems. But ISO's plan to make the OSI protocols a single standard for all computer users was defeated by ISO's own practice of approving multiple standards. The efforts of the international standards bodies contributed, sometimes unintentionally, to a computing environment that was characterized by heterogeneity both within and among networks. In this environment, the Internet—designed to handle diversity at all levels—had a competitive advantage.

By the early 1980s, computer owners who wished to network their machines had an increasing (and potentially confusing) number of options. The Department of Defense was working hard to promote TCP/IP. Ethernet and token-based systems were starting to become available for local-area networks. X.25 was a convenient choice for public network operators. Organizations that bought all their computers from a single manufacturer could take advantage of proprietary network systems. For interconnecting heterogeneous networks, however, there were really only two choices: OSI and TCP/IP. OSI was the official international standard, but it was years before protocols had been developed for all the layers in the model. TCP/IP, on the other hand, was readily available by the early 1980s and was backed by the expertise and experience of a large segment of the computer science community. TCP/IP reaped the benefits of its early start to become a de facto network standard both in the United States and, increasingly, abroad. At the same time, the Internet community succeeded in getting versions of TCP and IP sanctioned by ISO. This largely neutralized the standards controversy by paving the way for convergence—or at least accommodation—between the ARPA and OSI systems. Having survived the standards war, the Internet emerged with an even stronger basis for worldwide expansion.

The debate over network protocols illustrates how standards can be politics by other means. Whereas other government interventions into business and technology (such as safety regulations and antitrust actions) are readily seen as having political and social significance, technical standards are generally assumed to be socially neutral and therefore of little historical interest. But technical decisions can have far-reaching economic and social consequences, altering the balance of power between competing businesses or nations and constraining the freedom of users. Efforts to create formal standards bring system builders' private technical decisions into the public realm; in this way, standards battles can bring to light unspoken assumptions and conflicts of interest. The very passion with which stakeholders contest standards decisions should alert us to the deeper meanings beneath the nuts and bolts.

6

Popularizing the Internet

In the 1990s the Internet emerged as a public communications medium, and there were countless commentaries on its social impacts and implications. To the novice user, the Internet seemed to be an overnight sensation—a recent addition to the world of popular computing. The reality was different. In addition to the two decades of work that had gone into the development of packet switching networks, it took a series of transformations over the course of the 1980s and the early 1990s to turn the Internet into a popular form of communication.

At the beginning of the 1980s, the Internet included only a relatively small set of networks, most of which had direct links to defense research or operations. Over the course of the 1980s and the 1990s, the Internet would grow enormously in the number of networks, computers, and users it included; it would be transferred from military to civilian control; and its operation would be privatized, making the network much more accessible to the general public. Only then could most people grasp the possibilities for information gathering, social interaction, entertainment, and self-expression offered by the Internet and by an intriguing new application called the World Wide Web.

The question of who was responsible for creating this popularized Internet has no simple answer, because no single agent guided the system's evolution. ARPA was the original creator of the Internet technology, but during the 1980s that agency relinquished control over the Internet itself. A host of new actors assumed responsibility for various aspects of the system, including the National Science Foundation, the Bush and Clinton administrations, various public and private bodies outside the United States, university administrators, Internet service providers, computer vendors, and the system's many users.

With the loss of a central guiding vision from ARPA, the system seemed at times to verge on anarchy, as control of the network became fragmented among diverse groups with competing interests and visions. The Internet was also swept up in fast-moving changes in the technology, economics, and politics of computing and communications that made it difficult for anyone to foresee or plan its long-term development.

How did the Internet fare as well as it did under these turbulent conditions? I argue that the combination of an adaptable design and a committed user community accounts for its success. On the technical side, the Internet's modularity made it possible to change parts of the network without disrupting the whole, its robustness allowed it to function under the stress of rapid growth, its scalability helped it expand gracefully (although it did encounter some bottlenecks), and its ability to accommodate diversity allowed it to incorporate new types of networks. The techniques that made the Internet so adaptable—the TCP/IP protocols and the system of gateways—were adopted by network builders around the world, who hoped to join their networks to the Internet or at least achieve the same technical benefits. On the social side, ARPA (and later the NSF) worked hard to expand access to the Internet and to make TCP/IP easily available to potential users. The culture of the Internet also contributed to its widespread appeal. The Internet community's decentralized authority, its inclusive process for developing technical standards, and its tradition of user activism encouraged new groups to participate in expanding and improving the network, and the openness of the system invited users to create new applications (of which the World Wide Web would be the most dramatic example).

In this chapter I describe how the Internet was transformed from a research tool into a popular medium. I follow the system's growth and reorientation toward civilian research during the first half of the 1980s and the subsequent role of the National Science Foundation in further expanding it and, eventually, turning its operation over to the private sector. Along the way I explore some technical, managerial, and political issues raised by the Internet's expansion, privatization, and increasing economic importance. I then consider how the Internet became a global network, examining the role of independent networking developments in the United States and elsewhere and concluding with the emergence of the World Wide Web in the early 1990s.

Increasing Civilian Participation

At the start of the 1980s, the Internet—still under military control—consisted of a mixture of operational and research networks, many still experimental. Over the course of the 1980s, the balance shifted away from military involvement and toward academic research. New groups of researchers from outside the community of ARPA contractors began to gain access to the ARPANET, and military users moved to their own, more defense-oriented networks.

The first step in expanding civilian access to the Internet was initiated by the community most intimately involved with the technology: computer scientists. In the late 1970s, only a dozen or so computer science departments were connected to the ARPANET. The schools with ARPA contracts enjoyed access to specialized computers and increased professional communication and collaboration—benefits that their colleagues without access to the ARPANET noticed with envy. This created a demand for network access that had not existed before, as computer scientists at the majority of schools without ARPA contracts—for whom there were no comparable networking facilities—began to feel that they were at a professional disadvantage.

In May of 1979, Lawrence Landweber, chairman of the University of Wisconsin's computer science department, called a meeting of his colleagues at a number of schools to discuss possible solutions to their lack of network access. Beyond their local institutions' resources, university computer scientists had two main sources of funding: ARPA and the National Science Foundation. While ARPA tended to provide larger amounts of money, it supported only a select few research groups, whereas the NSF distributed its grants among a much larger number of schools. Landweber's group believed that the NSF would be receptive to the idea of funding a network that would serve a large number of researchers, and their efforts were encouraged by the head of the NSF's Computer Science Section, Kent Curtis. They submitted a proposal to the NSF for a new network, called CSNET, that would link computer science departments around the country.

This first proposal, which would have used public X.25 networks, was turned down by NSF reviewers on the basis of its technical design. In June of 1980 the group held a second planning meeting. This time Vint Cerf from ARPA attended the meeting, and he suggested some key changes in the CSNET plan. Cerf proposed that CSNET use the

Internet protocols, which had not been part of the original design, and he offered to set up connections between CSNET and ARPANET. Cerf's plan promised to benefit both ARPA-funded and non-ARPA-funded researchers by creating a single online community of computer scientists. And for ARPA there was another advantage. In 1980 TCP/IP was still being used by only a few sites, and ARPA managers were eager to get more people involved in using the new protocols. Creating a new network that used TCP/IP would ensure that the ARPA protocols got the attention and support of the computer science community (Landweber 1991).

The new plan for CSNET was further enhanced by an idea presented at the meeting by Dave Farber, a computer scientist at the University of Delaware. Farber described work he and his colleagues were doing on a system that would make it possible to build a low-cost network using dial-up telephone links. This offered a way to expand CSNET membership to schools that could not afford a full-time network connection.

After the meeting, the planning group prepared a second proposal for a composite network that would link sites by combining leased connections from the commercial Telenet network, a set of dial-up telephone connections that was referred to as PhoneNet, and the ARPANET. TCP/IP would be used by all the hosts communicating over this new internet. In 1981 the NSF granted $5 million to fund the CSNET project. To build their new network, the computer scientists created the PhoneNet system and set up Internet gateways between the ARPANET, Telenet, and PhoneNet. Computer science departments would connect to one of the three constituent networks, depending on their funding situation: ARPA contractors used the ARPANET, sites that could afford a full-time network connection subscribed to Telenet, and sites with little funding relied on PhoneNet. CSNET began operation in June of 1982 and was funded by the NSF until 1985, when it became self-supporting through member dues. The system included about 25 ARPANET hosts, 18 hosts using Telenet, and 128 hosts on PhoneNet (Quarterman and Hoskins 1986, p. 945).

The network created by the computer scientists broadened access to the Internet considerably. CSNET membership was open to any computer science institution—academic, commercial, nonprofit, or government—that was willing to pay dues. (Commercial use of the network was prohibited.) Putting a host on the ARPANET had required an expensive investment in hardware and software. CSNET's Telenet service was less costly, and the PhoneNet option offered network access that any school could afford.[1] CSNET also expanded international links to the Internet, at least for the limited purpose of exchanging electronic mail. CSNET was permitted to set up email gateways to the research networks that had been built in Germany, France, Japan, Korea, Finland, Sweden, Australia, Israel, and the United Kingdom. Those countries agreed to make sure that their links to the Internet were used only for approved research purposes (Landweber 1991; Quarterman and Hoskins 1986, p. 945).

While the Internet was still largely confined to the scientific community, CSNET set a precedent for opening up access beyond ARPA's own contractors. Internet managers Cerf and Kahn, who had come from the academic world themselves, were strongly disposed to expand network access among universities, and under their management the distinction between military and civilian use became something of a fiction. As long as the universities provided their own local infrastructure (Telenet or PhoneNet connections), and as long as the government's network was not exploited for profit, there were few political obstacles to opening up the system.

It became even easier to allow civilian access to the ARPANET in 1983, when the Department of Defense split the ARPANET into the MILNET (for military sites) and ARPANET (for civilian research sites). After the split, the civilian and military networks developed along separate paths. The Department of Defense continued to develop and operate a number of networks, both classified and unclassified, and the MILNET was incorporated into a larger system of military networks known as the Defense Data Network or the Defense Integrated Systems Network.

The purpose of the ARPANET/MILNET split had been to separate the military's operational and research communities so that they could manage their respective networks according to their own needs and priorities; there was no immediate plan to relinquish military control of the ARPANET. Still, the network split would make it more feasible to turn the Internet into a public service. Had the Department of Defense continued to use the ARPANET for its daily operations, it seems doubtful that the network would ever have been opened to the public. But the security restrictions that the Defense Communications Agency had imposed on the Internet during its administration were no longer a concern now that the military had its own networks. With university researchers once again the dominant population on the ARPANET, the Internet took on a decidedly more civilian character.

Growth at the Periphery

Another way that civilian researchers gained access to the Internet was through local-area networks at their universities. One of the most striking things about the Internet in the 1980s was its meteoric growth. In the fall of 1985 about 2000 computers had access to the Internet; by the end of 1987 there were almost 30,000, and by October of 1989 the number had grown to 159,000 (MERIT 1997). Most of the explosive growth of the Internet during the latter half of the 1980s came not from an expansion of the ARPANET itself but from the growing number of networks that were attached to it.

Where did these new networks come from? The growth of local-area networks was spurred by a computing revolution of the late 1970s and the 1980s: the rise of small, locally controlled computers. The first type of small computer to spread through the research community had been the minicomputer, introduced in the early 1960s, which allowed individual research groups to own and administer their own computers. The 1970s brought the even smaller and more affordable microcomputer—soon dubbed the "personal computer," since an individual could afford to own one. Unlike the large machines of the computing world, with their military and corporate roots, personal computers had sprung from the culture of amateur electronics hobbyists. The first personal computer, the Altair 8800, had been introduced in 1975 as a build-it-yourself kit.[2] By 1977 there were a number of "plug-and-play" machines available in the United States, including the Apple II, the Commodore PET, and the Tandy/Radio Shack TRS-80. In 1981 IBM entered the market with its own PC, which quickly became an industry standard. Personal computers were marketed as machines for lay people rather than experts and for use in the home as well as the office or laboratory. Many Americans were eager to own a computer, whether they were fascinated by the technology itself or whether they were hoping to realize the gains in skill and productivity that it promised.

In the 1980s, a new type of single-user minicomputer, the workstation, was adopted by many corporations and academic institutions. Workstations typically featured the Unix operating system and sophisticated graphical displays. As microcomputer technology evolved, the distinctions between personal computers and workstations began to blur and they began to share much of the same underlying hardware. The growing popularity of single-user computers in universities and businesses stimulated a demand for local-area networks to connect them. When research teams had shared the use of large time sharing computers, they could send email or share data by moving files around within a single computer. Once they were using separate computers, it became harder to share information—unless those computers could be networked together. Xerox PARC researcher Robert Metcalfe addressed that very need in 1975 when he devised the Ethernet system, which provided a simple and inexpensive way to network computers within a local area. Metcalfe later left Xerox to form a company called 3Com to commercialize his invention, and in the early 1980s 3Com introduced commercial Ethernet products that made it easy for people to build their own LANs for Unix workstations and personal computers. Ethernet was eagerly adopted by organizations with large numbers of small computers, and by the mid 1990s there were 5 million Ethernet LANs in operation (Cerf 1993; Metcalfe 1996, p. xix). Other LAN technologies, such as token ring and token bus, were also introduced. Universities and businesses quickly began building LANs, and those that had ARPANET connections began attaching their LANs to the Internet.

Whereas the growth of the ARPANET had been centrally planned, the attachment of LANs to the Internet was a remarkably decentralized phenomenon, depending largely on local decisions at the individual sites. The modularity of the Internet made it relatively simple to attach new networks—even those that used a very different technical design, such as Ethernet. ARPA managers Cerf and Kahn permitted and encouraged contract sites to connect their LANs to the Internet. This would have been a rather extraordinary move for a commercial network; however, ARPA was not in the business of selling Internet service, so its managers had no incentive to restrict access for economic purposes. From their perspective, having a larger user community enhanced the value of the Internet as a research tool with little extra cost to the agency, and the robust and decentralized nature of the system minimized the need for ARPA to exercise central control over its expansion. That few outside the research community knew or cared about the Internet in the early 1980s also helped make it politically feasible for ARPA to let the system expand in an informal way. No one in Congress was arguing over who should or should not be allowed access to the Internet, or at what cost.

In fact, far from restricting access, ARPA took an active part in making it easier for sites to create TCP/IP-based LANs and connect them to the Internet. In order for an ARPANET site's local network to be connected to the Internet, two technical requirements had to be met: the site had to run TCP/IP on the local network, and it had to set up a gateway (also called a router) between its network and the ARPANET. ARPA helped out with both tasks. Dave Clark at MIT, who coordinated the technical development of TCP/IP through most of the 1980s, provided versions of TCP/IP that could run on personal computers (Leiner et al. 1997). ARPA funded a number of vendors to develop TCP/IP products for Ethernet, and it published its own official standard for transmitting IP packets over Ethernet in April of 1984; the companies involved often went on to commercialize these products (Hornig 1984). By 1985 there was a healthy commercial market for products that allowed minicomputers and microcomputers to run TCP/IP over Ethernet. ARPA contractors also developed and commercialized Internet routers, and by the mid 1980s a market had developed for off-the-shelf routers.

Thus a series of developments that began far from ARPA came to have a significant impact on the Internet. The combined effect of the growth of PCs and LANs, the commercial availability of TCP/IP software and routers, and ARPA's open-door policy was that LANs began joining the Internet in droves (Cerf 1993). In 1982 there had been only 15 networks in the Internet; four years later there were more than 400 (NSF Network Technical Advisory Group 1986, p. 3). The addition of LANs to the Internet meant that a new group of local network managers took on responsibility for managing parts of the system. As universities attached their LANs to the Internet, its resources became accessible to academics who were at ARPA-funded institutions but were not necessarily involved in work funded by or related to the military.

What's in a Name?

Although the Internet could afford to grow in a decentralized and spontaneous way, there were still certain functions for which central coordination seemed to be needed to prevent chaos. One of the most important of these functions was providing a uniform and comprehensive system of host names and addresses that would allow each computer to be uniquely identified. In order for Internet hosts to exchange messages, each host had to be able to obtain the addresses of all the others; there also had to be some way to make sure that no names or addresses were duplicated. This called for some type of system-wide coordination.

In the early 1980s, as the number of hosts and networks on the Internet began to rise, the original naming system showed signs of strain. Each host computer on the Internet has both a *name* (a set of characters—often a recognizable word—that can be used to refer to the host) and a numerical *address* (used by the network to identify the host). This dual identity relieves users of having to deal with cumbersome numerical addresses, but the system requires that there be a way to map names onto addresses. To translate names into addresses in the old Internet system, each host kept a table of names and addresses for all the other hosts on the Internet. The host table had to be updated whenever hosts were added, were removed, or changed their point of attachment to the network—events that occurred often. The Network Information Center was responsible for approving new host names and for maintaining and distributing updated host tables; however, it often fell behind on these tasks, and many host administrators began using their own unapproved but up-to-date host tables.[3] When the NIC did distribute updated host tables, the sheer size of the files threatened to swamp the network with traffic as they were sent out to hundreds of hosts. Host administrators complained about the inadequacy of this system, and it was clear to everyone that further growth of the Internet would only exacerbate the problem.

To address this issue, members of the Internet technical community began discussing the idea of dividing the Internet name space into a set of smaller "domains." Host names would take the form "host. domain," and individual users would be identified as "user@host. domain." The Domain Name System, as it came to be called, was largely designed by Paul Mockapetris at the University of Southern California Information Sciences Institute, and was adopted by the Internet in the mid 1980s (Cerf 1993; Leiner et al. 1997).[4] Its goal was to distribute the task of maintaining host information in order to make it more manageable. Instead of having a single organization maintain files of all the host names and addresses, each domain would have at least one "name server," a special host that maintained a database of all the host names and addresses within that domain. When a host needed to find the address of another host, it would send a query to the name server for the destination host's domain, and the name server would return the address of the destination host.[5]

The Domain Name System eliminated the need to distribute large files containing host tables across the network at frequent intervals. Instead, updated host information would be maintained at the name servers for the various domains. Host computers would no longer have to keep tables listing hundreds of host names and addresses; now they needed to know only the addresses of a small number of domain name servers. Later, the system was adapted to recognize addresses on non-TCP/IP networks, which made it easier for people whose networks were not part of the Internet to exchange mail with Internet users.[6]

Domains could theoretically represent any subset of the Internet, such as an organization, a type of organization, or even a random selection of hosts. In practice, ARPA decided to create six large domains to represent different types of network sites: edu (educational), gov (government), mil (military), com (commercial), org (other organizations), and net (network resources). (Additional domains were subsequently added.) This division by type of host was designed to make it easier to manage the domains separately: the military could control the "mil" domain, an educational consortium could administer the "edu" domain, and so on.[7] Beneath the top-level domains were other, site-specific domains, and these in turn could be further divided to create a nested hierarchy of domains. For instance, within the top-level domain "edu," each university would have its own domain; a university could then choose to give different departments or other groups their own domains within the university domain (Mills 1981; Krol 1992, pp. 26–27). This decentralized the naming process: the administrator of each domain could assign lower-level domain names without consulting a central authority. Host names had to be unique within a domain, but the same host name could be used in different domains, since the *combination* of host and domain names would still be unique. If people at several universities wanted to name a machine "frodo" after the character in J. R. R. Tolkien's *Lord of the Rings* (a common preference among computer science students), it would be permissible. Thus, there was no need to coordinate naming above the level of the local domain. The Domain Name System provided a way to keep the task of finding addresses manageable, to facilitate email exchange among diverse networks, and to distribute the authority for naming hosts and lower-level domains (Mills 1981; Krol, 1992, p. 27; Cerf 1993).

The National Science Foundation Takes the Lead

ARPA's activities in funding computer science during the 1960s and the 1970s had been paralleled, on a much smaller scale, by those of the National Science Foundation. In the 1980s, in the wake of its sponsorship of CSNET, the NSF began to build large networks of its own and became involved in the operation of the Internet. The NSF's pursuit of networking greatly expanded the size and scope of the Internet, opened up access to virtually every interested university, and eventually brought the Internet under civilian control.

The NSF's Office of Computing Activities had supported computing centers at various universities since the mid 1960s, and NSF managers were sympathetic to the idea of building networks to connect these centers (Aufenkamp and Weiss 1972, pp. 227–228). As early as 1972 the NSF began studying the possibility of developing a national network to pool hardware resources, encourage collaboration beyond institutional boundaries, and share programs and databases (ibid., p. 226). The NSF did not attempt to build such an ambitious network in the 1970s, perhaps because of its relatively small budget and perhaps because the potential importance of such a network to researchers was not yet recognized. Instead, it funded some smaller networks that linked clusters of universities on a regional basis.

In the 1980s, the NSF began planning its own nationwide network. The impetus for this was the NSF's new supercomputer program. In mid 1984 the NSF created an Office of Advanced Scientific Computing, whose mandate was to establish several new supercomputer centers around the United States. To make those publicly funded machines available to a wider community of researchers, the NSF simultaneously began planning for a high-speed network to link the supercomputer centers and provide access to other universities (Quarterman and Hoskins 1986, p. 309). The NSF would spend $200 million to operate the NSFNET over the next ten years.

The NSF planned its network as a two-tier system. University computer centers would be linked to regional networks, which would be connected in turn to a central network known as the "backbone" of the NSFNET. The backbone would comprise a set of packet switches connected by high-speed leased lines. Each participating regional network or supercomputing site would have a gateway to one of the backbone switches. The first version of the backbone linked the NSF's six supercomputer sites; eventually, the backbone included 16 nodes, each of which served one or more supercomputer centers, national laboratories, or regional networks (Wolff 1991, p. 1). Thus, the NSFNET was conceived from the beginning as an internet, not a single network.

The idea of building regional networks followed from earlier NSF activities. In the period 1968–1973, the NSF's Office of Computing Activities had funded 30 regional computing centers as a way to help universities make efficient use of scarce computer resources—and also to make sure that elite schools would not be the only ones to benefit from computers (Aspray 1994, p. 69). One offshoot of this program was a number of networking experiments aimed at making access to these regional centers easier. The NSF subsidized regional networks that enabled cooperating institutions to share resources; it also provided startup funding to help academic consortia build their own self-supporting networks. Examples of consortia-run networks included the New England Regional Computing Program (NERComp), which built a network connecting 40 New England universities to seven main computing sites in 1971; the Michigan Educational Research Information Triad (MERIT) network, begun in 1972; and EDUNET, a nationwide education network built by EDUCOM, a consortium of educational computer centers and networks (Cornew and Morse 1975, p. 523; Farber 1972, p. 38; Quarterman 1990, p. 318; Emery 1976).

With the inauguration of the NSFNET project, the NSF stepped up its support for regional networks. With encouragement from NSF program managers, groups of universities in various parts of the United States organized regional projects and submitted funding proposals to the agency. By early 1988, seven new regional networks were in operation, including BARRNet (in the San Francisco Bay area), MIDNet (in the Midwest), NorthWestNet, NYSERNet (in the New York area), Sesquinet (in Texas), SURAnet (in the Southeast), and WESTNET (in the Rocky Mountain region) (Quarterman 1990). In addition, the NSF sponsored a number of regional networks specifically for supercomputer access and the University Satellite Network (USAN), which linked several universities by satellite to the National Center for Atmospheric Research in Boulder. These new systems and the existing MERIT network were all linked to the NSFNET by 1988 (figure 6.1). Other regional networks continued to be created and linked to the NSFNET backbone.[8]

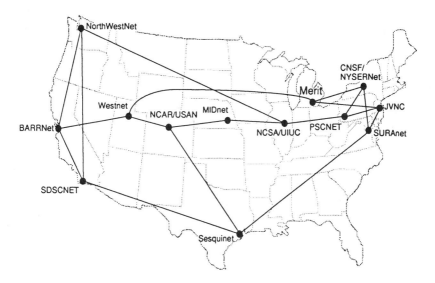

Figure 6.1
The NSFNET in 1989. Source: MERIT 1989. The supercomputer networks were JVNCNet at Princeton, NCSAnet at the National Center for Supercomputing Applications at the University of Illinois, PCSnet at the Pittsburgh Supercomputer Center, and SDSCnet at the San Diego Supercomputer Center.

The initial version of the NSFNET, designed collaboratively by personnel at the various sites, was only temporary (MERIT 1995). The NSF sought competitive bids from commercial contractors to build a more advanced version of the backbone, and in 1987 it awarded a five-year contract for building and operating an upgraded version to MERIT, with IBM to supply the packet switches and MCI to provide the leased lines.[9] The original network had used the "fuzzball" protocols created by David Mills of the University of Delaware, but the designers of the upgraded network, led by Dennis Jennings of the NSF, decided that it should use TCP/IP. This move reflected an effort by the NSF and ARPA to coordinate their network activities. To pool their resources, the two agencies agreed that while the new NSFNET backbone was being designed and built the NSFNET would use the ARPANET as its backbone and the NSF would share some of the ARPANET's operating costs (Barry Leiner, email to author, 29 June 1998). This scheme required that the NSF's regional networks run TCP/IP so as to be able to communicate with the ARPANET.

The NSF-ARPA interconnection arrangement opened the Internet to nearly all the universities in the United States, making it a civilian

n all but name. The NSFNET also introduced some new
the Internet system, including nonprofit regional network
operators and the new supercomputer centers (which provided the
network's most impressive resources and created concentrations of
computer experts around these resources). The regional networks and
the supercomputer centers would play important roles in the privatization and popularization of the Internet.

The End of the ARPANET

As the Internet grew, its backbone network, the ARPANET, was beginning to show its age. By the late 1980s the ARPANET was almost 20
years old—quite a long time in the computer field. The system's IMPs
and 56-kilobits-per-second lines no longer had the capacity to serve
the escalating number of users of the Internet system, which was
estimated in 1987 to have several hundred thousand computers and
as many as a million users.[10] In December of 1987, the managers of
ARPA's network program, Army Major John Mark Pullen and Air
Force Major Brian Boesch, decided that the ARPANET had become
obsolete and would have to be retired.

But what would replace the ARPANET as the communications system for ARPA researchers and the backbone of the Internet? At first,
the ARPA managers envisioned building a new network (to be called
the Defense Research Internet), transferring users to it, and then
dismantling the ARPANET. But there was another possibility: to connect the ARPANET sites to the NSF's regional networks, and to have
the NSFNET take over as the backbone of the Internet. The NSFNET
was being designed with higher-speed lines and faster switches than
the ARPANET, so it would be able to handle more traffic. Since the
NSF and ARPA were already operating their network services jointly,
and since many of their sites overlapped, this option appealed to the
ARPA and NSF managers. When the NSFNET backbone was ready, it
would simply be a matter of transferring the entire Internet community from the ARPANET to the NSFNET.

The design choices that Cerf and Kahn had made in creating
TCP/IP made this backbone swap relatively easy. The Internet's
designers had decided to give the host computers, rather than the
network itself, responsibility for most of the complicated networking
functions. Those functions would not be disrupted, therefore, by
changes in the backbone networks. Keeping the network's tasks simple
had given the Internet system needed flexibility.

During 1988 and 1989, the various ARPA contract sites transferred their host connections from the ARPANET to the NSFNET. On 28 February 1990 the ARPANET was formally decommissioned and the remaining hardware dismantled. The changeover caused little disruption in network service; most ARPANET users were probably not aware that the transition had taken place. But Vint Cerf penned a "Requiem for the ARPANET," which concluded as follows:

It was the first, and being first, was best,
but now we lay it down to ever rest.
Now pause with me a moment, shed some tears.
For auld lang syne, for love, for years and years
of faithful service, duty done, I weep.
Lay down thy packet, now, O friend, and sleep.
(Cerf 1989)

Cerf's sentiments were echoed by other ARPANET veterans who had witnessed the network's remarkable 20-year evolution from an uncertain experiment to a system that routinely served hundreds of thousands of users. The end of the ARPANET was not simply a sentimental occasion, however: the passing of the baton from ARPA to the NSF also marked the end of military operation of the Internet.

Privatization

Although the Internet had come under civilian control, it was still run by a government agency and still intended only for nonprofit research and education. The final step toward opening the network to all users and activities would be privatization. The issues that the NSF faced in trying to privatize the Internet were in some ways very characteristic of US attitudes toward the role of the federal government. Americans tend to disapprove of government involvement in providing commercial goods or services, as the heated debates in the 1990s over the establishment of a national health care system or federal subsidies for high-tech research and development illustrate. Therefore, the NSF managers believed that the only politically feasible way to accommodate commercial users on the Internet would be to remove it entirely from government operation.

ARPA managers had tried as early as 1972 to persuade a commercial operator such as AT&T to take over the ARPANET, but they had not been successful; in 1972 it was not evident that the market for data network services was big enough to interest a giant corporation. By the early 1990s, however, the Internet had grown by several orders of magnitude, the advent of personal computing had vastly expanded the potential market for network services, and the once-monolithic telecommunications industry had been opened to smaller carriers who might be more interested in entering the computer networking market. The National Science Foundation would find privatizing the network more feasible than ARPA had, though the task still raised some thorny issues.

At the beginning of the 1990s the NSF had to make difficult decisions about the future of the Internet, some having to do with the network's users and others with the contractors who operated it. All users of the NSFNET backbone (which was now the Internet backbone too) were required to abide by the NSF's Acceptable Use Policy, which stated that the backbone was reserved for "open research and education" and which specifically prohibited commercial activities.[11] The Acceptable Use Policy was a political necessity, since Congress was quick to condemn any use of government-subsidized resources for commercial purposes.[12] However, the policy was unpopular with users, and the fact that many sites were involved in both research and commercial activities made it hard to enforce. It was clear to members of the Internet community that the conflict between policy and practice would only get worse as more and more businesses began using computer networks.

At the same time, commercial network service providers were demanding that the NSF give them an opportunity to compete for the business of providing backbone services. MERIT and its partners IBM and MCI were operating the backbone under a contract from the NSF that would expire in 1992. In 1990, this group spun off a nonprofit corporation called Advanced Network Services (ANS) and subcontracted the backbone operations to this new entity (MERIT 1995; Wolff 1991). ANS then set up its own for-profit counterpart, which began offering commercial network services. These developments caused consternation within the Internet community: it was one thing to have a nonprofit consortium such as MERIT running the network, but quite another to have a commercially involved enterprise such as ANS exercise monopoly control over the provision of Internet backbone services.

NSF managers saw privatization as the solution to their worries about users and contractors. If they could shift the operation of the Internet from the NSF to the commercial sector and end direct government subsidies of its infrastructure, the issue of acceptable use would disappear. And with the private sector supplying Internet services, network companies could compete for customers in the marketplace, rather than competing for NSF contracts. In 1990, NSF manager Stephen Wolff began discussing the idea of privatizing the Internet with interested members of the Internet community, holding workshops and soliciting comments from network experts, educational groups, and representatives of other government agencies. Wolff found a "broad consensus" within the Internet community that the NSF should arrange for several competing companies to provide backbone services (Wolff 1991, pp. 1–2).[13] The question was how to plan the transition from government to private operation in a way that was both equitable and technically feasible.

One development that aided the NSF's efforts was the rise of commercial network services. In the short time since the NSFNET had been created, the American networking environment had changed dramatically. In 1987, the operator of the ARPANET—Bolt, Beranek and Newman—had been the only company with experience running a large-scale TCP/IP network. But the subsequent growth of the NSFNET spawned a number of commercial network service providers, and by 1991 there was a thriving and competitive market for high-speed nationwide computer networking services. The possibility existed, therefore, of having several companies share responsibility for the Internet backbone.

Where had all these commercial networks come from in so short a time? Some of them were direct spinoffs of the NSF's own regional networks. Just as members of the ARPANET project had spun off commercial services such as Telenet, so the creators of regional networks had created commercial Internet ventures. The first of these entrepreneurs was William L. Schrader, who had led the creation of NYSERNet in 1986. Like all the NSF regional networks, NYSERNet used a physical infrastructure of communications links and packet switching computers to provide a particular service: TCP/IP-based connections between the region's host computers and the rest of the Internet. What Schrader and the other network operators who followed him did was separate—conceptually and legally—the operation of the infrastructure from the provision of the service.

By the late 1980s, the use of computers and local networks in business had grown to the point where a potentially lucrative market existed for networking services. In 1989 Schrader founded Performance Systems International (later known as PSINet) and began offering TCP/IP network services to business customers (PSINet 1997).[14] To provide these services, PSINet bought out NYSERNet's infrastructure. NYSERNet became a broker of network service rather than a provider, buying service from PSINet and selling it to NSF-sponsored users. Since the network infrastructure was no longer directly paid for by the US government, PSINet could also sell its services to business customers for additional profits. PSINet proved to be a successful business venture, and other spinoffs of regional networks quickly followed along the same lines.

The new network providers served both NSF research and education sites and commercial customers. However, since commercial traffic was still forbidden on the NSFNET, only the NSF-sponsored sites could send traffic through the NSFNET backbone; traffic from commercial customers had to be routed through the network service provider's own backbone. This encouraged most of the new companies to expand their backbone operations from their original regional areas to the entire continental United States. MCI, AT&T, Sprint, and other telecommunications carriers also began to offer commercial Internet services (Cerf 1993). By the mid 1990s, a whole parallel structure of commercial TCP/IP networks had evolved.

One handicap for these service providers was that the only connection between their various networks was the Internet backbone, which was off limits to traffic from commercial customers. To increase the scope of service for commercial users, three of the new service providers—PSINet, CERFNet, and Alternet—joined together in July of 1991 to form a nonprofit organization called the Commercial Internet Exchange (CIX). The CIX set up a gateway to link the three networks, the operation of which was financed by a membership fee, and the members agreed to accept traffic from any other member network free of charge. This free-exchange policy spared CIX members the considerable trouble and expense of setting up the technology to support an accounting system for network traffic. In any case, the network providers would have found it difficult to pass on access charges to their customers. Unlike the telephone system, computer network customers were not charged on the basis of how far their packets traveled; indeed, they often did not even know the physical locations of the computers to which they sent packets, and they were even less likely to know which commercial network a computer was on. Under these circumstances, trying to impose charges for sending packets between networks would have caused great technical difficulties and, in all likelihood, would have outraged customers.

The CIX arrangement, which allowed the customers of any member network to reach users on all the other networks, greatly increased the value of the service each network provided. Other commercial networks soon joined the CIX to gain these benefits for themselves, and there were eventually dozens of CIX members worldwide. Since the commercial networks were also providing the Internet's regional infrastructure, the only physical difference between the CIX and the Internet was that they had different backbones. With the commercial networks imposing no restrictions on the type of traffic they would carry, the CIX became, in effect, a commercial version of the Internet, offering the same set of connections to a different clientele.

With this commercial infrastructure evolving rapidly, the NSF could plan to replace its own Internet backbone with a commercially based operation. In November of 1991 the NSF issued a new Project Development Plan, which was implemented in 1994. Under the new plan, Internet service would be taken over by competitive Internet Service Providers (ISPs), each of which would operate its own backbone, and the old NSFNET backbone would be dismantled.[15] Customers would connect their computers or LANs to one of the commercial backbones. There would be a set of gateways, called "exchanges," at which two or more ISPs would connect their systems according to bilateral agreements, thus allowing traffic to be sent from one network to another.[16]

The government, meanwhile, would create a new segment of the Internet, called the "very-high-speed Backbone Network Service" (vBNS), which would be restricted to specialized scientific research. The NSF gave contracts to four Internet Service Providers to operate a set of gateways between their networks and the vBNS, so as to ensure easy access for the research system's users.[17] Aside from the research-oriented vBNS, however, the commercial version of the Internet had become the only version. On 30 April 1995, MERIT formally terminated the old NSFNET backbone, ending US government ownership of the Internet's infrastructure (MERIT 1995).

With privatization, the Internet was opened up to a much larger segment of the American public, and using it for purely commercial, social, or recreational activities became acceptable. Commercial online services could now offer Internet connections, and the computer industry rushed into the Internet market with an array of new software products and services. Corporations that had built their own long-distance data networks because of the prohibition on commercial use of the Internet could put their computers on the Internet and phase out their expensive private networks (Krol 1992, p. 17). As a flood of new users joined the network, the Internet suddenly became the focus of new social issues involving personal privacy, intellectual property, censorship, and indecency. At the same time, network users created a whole new set of applications (for example, the role-playing games known as "multi-user dungeons") to fulfill their desires for entertainment, social interaction, and self-expression. The Internet became a topic of public discussion, and ordinary people began to debate the advantages and pitfalls of "going online."

Convergence with Other Networks

In parallel with the development of the Internet in the 1970s and the 1980s, a host of other networks with diverse technical approaches, management philosophies, and purposes had been created. In the days when Internet access was restricted, these networks provided alternative options for network service. With the privatization of the Internet, the independent networks contributed a large population of experienced network users to the Internet community, as well as some new applications that would appeal to a public in search of social interaction or amusement on the Internet.

The ARPANET had publicized the benefits of computer networking in the early 1970s. Later in that decade, a number of individuals and organizations began to experiment with providing these benefits to computer users who were excluded from the ARPA community and could not afford commercial network services. These grassroots networks, designed to be inexpensive, were usually run as cooperatives, with a minimum of central coordination. They were user-driven efforts; some received modest funding from the computer industry, but others had no outside support at all.

Some resourceful computer users improvised their own electronic message services, using operating system software that had been provided for other tasks. For example, when AT&T distributed the 1978 version of its Unix time sharing system, it included a program called UUCP (Unix-to-Unix copy) that allowed users to copy files from one computer to another. Almost immediately, computer users at universities, where Unix was widespread, took advantage of this program to create an informal email network. The UUCP network was a simple affair: instead of having its own networking infrastructure, it 1 periodic dial-up telephone connections between hosts to e mail files. The only central management was a map of participating hosts that was used to route mail through the system, and this was maintained by volunteers.

The UUCP software was also used by two students at Duke University, Tom Truscott and Steve Bellovin, as the basis of a system for the distribution of electronic newsletters. In 1979 they set up a news exchange system between Duke and the University of North Carolina, using dial-up connections. Word of this system spread to people at other universities, who were invited to copy the software and join the news exchange. Soon an informal network that came to be known as USENET was created. Described by its founders as "a poor man's ARPANET," USENET provided inexpensive network communications for many schools that had no other access to a national network (Quarterman 1990, p. 243). USENET was used to distribute online forums called "newsgroups" that featured a variety of different topics. Any user at a USENET site could submit messages to a newsgroup, which would then be available to all other readers of the newsgroup; this enabled users to participate in an ongoing discussion. Users could create newsgroups on any topic they wanted to discuss. Most of the early newsgroups focused on practical matters of using and operating computers, but soon social and recreational groups sprang up to discuss sex, science fiction, cooking, and other subjects. Computer users flocked to USENET because it offered new possibilities for social interaction, bringing together "communities of interest" whose members might be geographically dispersed and allowing people to participate anonymously if they chose. Users could select which newsgroups to read, and a number of programmers developed and distributed software that made it more convenient to select newsgroups and read messages. Designed and managed by its users and having no obligations to the government, USENET was even more decentralized and freewheeling than the Internet.

Another improvised service made use of the IBM RJE (remote job entry) protocol, a standard feature of the operating systems of IBM machines that allowed the user of one computer to submit programming jobs to another. Since the RJE software was designed to transfer program files from one computer to another, it only took a little modification to use it to exchange other types of files, such as mail. In 1981, Ira Fuchs at the City University of New York and Greydon

Freeman at Yale University obtained funding from IBM to create an experimental connection between their two schools using RJE. This became the first step in building a network for IBM users, which was named BITNET. (The "BIT" acronym stands for "Because It's There," referring to the system builders' adaptation of an existing protocol.) Like the UUCP network, BITNET was used primarily for electronic mail, although it also allowed a pair of users to "chat" in real time over a dial-up connection. IBM provided funding to support software development and administrative work at CUNY, which served as a hub for the exchange of mail among the various BITNET sites. This funding ended in 1986, and the network became self-support-ing through modest user fees. Like the ARPANET, BITNET and USENET were examples of how network users could take tools that had been designed for computation and adapt them for personal communication.

When personal computers became common, in the early 1980s, computer ownership became possible for a new groups of users who often did not have access to institution-based networks. Some users set up their computers to serve as "bulletin boards" by adding a modem and software that allowed others to dial in to the machine and post messages, to which other users could respond. One popular system for bulletin boards was named Fido. In 1983, Tom Jennings, the operator of a Fido bulletin board, created FidoNet, which used dial-up connec-tions to allow the exchange of messages between Fido machines. By 1990, when about 2500 computers had joined FidoNet, its applications ranged from a forum for handicapped people to a directory of data-bases created by the United Nations (Quarterman 1990, pp. 257–258).

Cooperative networks were organized in an informal way; joining a network required only that one arrange to periodically call another site on the network to exchange mail or news files. The expenses entailed in joining such networks were limited to the cost of one's telephone calls and sometimes a small membership fee; this made them attractive to individuals and organizations with limited comput-ing budgets, including political and social activists. Soon USENET, BITNET, and FidoNet were serving thousands of host computers in the Americas, in Europe, in Australia, in Asia, and in Africa (Quarter-man 1990, pp. 230–239). BITNET spawned branches in Canada (called NetNorth) and Europe (called EARN), all of which were con-nected to form a unified email system. Networks combining USENET and UUCP services were set up in Europe (EUnet) and in Japan

(JUNET). USENET, BITNET, and FidoNet also set up gateways to allow the exchange of mail with one another. These low-cost networks helped spread the benefits of network technology—previously the domain of the wealthier nations, organizations, and individuals—to less privileged groups.

By the mid 1980s, the grassroots systems were being imitated by commercial email services on existing telephone or computer network systems: MCI Mail, AT&T Mail, Telenet's Telemail, DEC's EasyLink, and others. In addition, IBM, DEC, and other large computer companies had built their own wide-area networks that linked the companies' employees around the world. These networks used the manufacturer's own proprietary protocols and were not open to outside users, so they did not represent an option for the general public, but they increased the overall population of network users who might later choose to join the Internet.

Another popular form of computer communication was the use of conferencing systems, also known as "online services." A conferencing system was not a network per se but rather a single computer site into which users could dial to post messages, download files, exchange email, or participate in real-time online conversations. The early 1980s saw the introduction of commercial online systems, such as Compu-Serve, America Online, and Prodigy, which catered to the personal computer user. Subscribers would access these services by means of a modem and software supplied by the service provider. In their original form, these online services did not offer access to the Internet (which was still restricted); they simply connected users to the provider's own computer system, which offered features such as free software, access to online shopping or other services, and the opportunity to "chat" with other subscribers. This non-networked form of online service, now largely forgotten, was instrumental in introducing large numbers of users to the practice of accessing information and interacting with other people via a distant computer.

Some conferencing systems were set up to serve particular communities or regions. In 1985, for instance, the WELL (Whole Earth 'Lectronic Link) was set up by Stewart Brand (of *Whole Earth Catalog* fame) and Larry Brilliant (head of a California software company) as an alternative to the commercial online systems (Figallo 1995, p. 51). The WELL, intended to foster a sense of local community for its members in the San Francisco area, became known as a gathering place for advocates of counterculture ideas and free speech. It was run on a modest commercial basis, charging minimal fees to recover its expenses and relying heavily on volunteer work by its users. Many other "alternative" conferencing systems—such as PeaceNet, created in 1985 by peace activists—were nonprofit services. Most were open to all interested participants for the cost of the telephone connection plus a nominal fee to cover operating expenses.

By the late 1980s, therefore, several million computer users could exchange mail and news over the various grassroots and commercial networks. Though these systems were not parts of the Internet, they established links to it fairly soon. ARPA permitted mail from other systems to be sent to Internet users, and in the early 1980s it sponsored the development of software that would allow hosts to act as "mail relays," receiving mail from one network and sending it into another after performing any necessary reformatting.[18] Sites that had connections to both USENET and the Internet began sending USENET news files between the two networks, and news soon became a standard feature for Internet users—so much so that in 1986 members of the Internet community developed a protocol called NNTP for the specific purpose of transmitting news files over TCP/IP-based networks.

The cooperative networks had been designed to use software that was specific to a particular type of computer or operating system, since their creators had used existing software tools. Eventually, people adapted the software for use on computers of other types; however, the various networks were still incompatible. In addition, because the networks used different naming and addressing schemes, users who wanted to send messages from one network to another were forced to use awkward address formats, such as[19]

uucp:host1!psuvax1!host.bitnet!username

or

username%host.domain.junetdomain@csnet-relay.csnet.

As more people joined these networks and had trouble communicating with users of other networks, the network coordinators began to consider adopting more general-purpose protocol standards. Outside the United States, network operators most often chose to adopt the OSI protocols, which were an international standard rather than an American one (Quarterman 1990, pp. 232, 252). In the United States, the cooperative networks chose to adopt the Internet protocols, both because they wanted better access to the large Internet community and because they considered TCP/IP to be the best available option for providing a common language between networks. In the late 1980s the US portion of BITNET gave up the RJE system in favor of TCP/IP, and a version of UUCP was developed to run over TCP/IP. In 1986 the BITNET and UUCP organizations also agreed to adopt the Internet's Domain Name System; FidoNet followed in 1988 (ibid., pp. 111, 256).

Even before privatization, then, the commercial and nonprofit networks had interacted extensively with the Internet. Once the Internet had been privatized, many users of cooperative networks began to switch to Internet Service Providers. This represented the convergence of two strands of network development: the users of grassroots networks adopted the Internet infrastructure, while the Internet community adopted newsgroups and other applications that had been popularized by the cooperative networks. Commercial online services also joined the Internet. In many cases they became little more than Internet service providers, abandoning their original role as content providers; however, chat rooms and other services they had popularized entered the mainstream Internet culture. These independently developed applications and the many users who had been drawn to networking through the grassroots and commercial services helped fuel the Internet's growth and popularity.

Management Issues

Privatizing the Internet backbone had been relatively easy, but the transition to commercial operation left open the question of who would provide ongoing technical planning and administration for the system. Each member network, from the smallest LAN to the largest Internet Service Provider, was responsible for its own operations. However, protocol development, administration of Internet names and addresses, and other tasks that affected the entire system still required some central coordination—a function that the National Science Foundation could no longer provide.

The NSF adopted a range of methods for delegating various aspects of the management of the system, many of them similar to approaches used in other newly privatized industries. Some coordination functions were vested in nonprofit, non-government bodies. The educational consortium MERIT, for instance, continued to act as the central authority on Internet routing information even after it ceased to operate the NSFNET backbone. In other cases, control over administrative functions was split among competing commercial entities, on the theory that market competition would encourage innovation and prevent any one interest group from gaining too much power. For instance, in the late 1990s, as Internet domain names such as "microsoft.com" began to be seen as valuable symbols of organizational identity and even intellectual property, the question of who should assign these names became hotly contested. Some people believed that name registration should remain under the central control of the InterNIC, a government-designated nonprofit body. Others, claiming that the InterNIC was slow, unresponsive, and careless in its business practices and arguing that competition would provide better service to people applying for domain names, proposed turning the job over to a number of commercial name registrars. This proved a difficult issue to resolve: since there might be several groups trying to register a particular name, there were many potential losers in the naming process, and they were likely to criticize whatever system was in place.

The technical side of managing the privatized Internet was little changed from the ARPANET days, in part because there was a good deal of continuity in the core group of computer experts who made the technical decisions. The ARPANET Network Working Group had set the style for technical development with its informal, participatory process and its use of Requests For Comments to propose and comment on protocol standards. This approach was similar to the committee-run, consensus-based style of many standards organizations, though without the usual restrictions on membership. The NWG disbanded in the early 1970s as the ARPANET went into full operation, but its methods were perpetuated by the series of technical oversight groups that succeeded it.

During the 1970s, the Internet Program—run by NWG alumni Vinton Cerf and Robert Kahn—took on responsibility for ongoing protocol development. Cerf and Kahn set up an advisory group of network experts called the Internet Configuration Control Board. One of this board's jobs was to encourage a wide spectrum of potentially interested parties within the network community to contribute to and debate the merits of the system's evolving protocols. If consensus on a proposed protocol seemed to emerge, the ICCB would often arrange for a few Internet sites to create implementations of the protocol to see how it worked under actual use; if the protocol was tested successfully, the ICCB would declare it an official Internet standard.

Barry Leiner, as ARPA's network program manager from 1983 to 1985, reorganized this management structure in an effort to broaden participation in decisions about the network's design. He replaced the rather small circle of the ICCB with a more inclusive body called the Internet Activities Board, which was chaired by Dave Clark of MIT for many years.[20] The leadership of the IAB drew heavily on the research community that had built the ARPANET, with members from the Department of Defense, from MIT, from Bolt, Beranek and Newman, from the University of Southern California's Information Sciences Institute, and from the Corporation for National Research Initiatives (a networking think tank formed by Cerf and Kahn). However, membership in each of these groups was open to anyone, anywhere in the world, who had the time, interest, and technical knowledge to participate.

The Internet Activities Board became a forum for discussing all aspects of Internet policy, and its meetings became very popular in the networking community. By 1989, the number of people participating in the IAB had grown into the hundreds, and its leaders decided to divide its activities between an Internet Engineering Task Force (which would lead protocol development and address other immediate technical concerns) and an Internet Research Task Force (which would focus on long-range technical planning) (Postel and Reynolds 1984, pp. 1–2; Kahn 1990). Working groups within these task forces coordinated their activities through email, and the task forces held meetings several times a year. Standards for the Internet were set by consensus, after discussion among all interested parties and after proposed protocols had been tested in practice, and they continued to be published electronically in the form of Requests for Comments (Quarterman 1990, pp. 184–186).

When the National Science Foundation began its NSFNET project, it set up its own Network Technical Advisory Group, chaired by David Farber at the University of Delaware. After the managers of the NSF and ARPA decided to merge their networks, the NSF folded its technical group into ARPA's Internet Engineering Task Force. In the ensuing years, the IETF took on members from the Department of Energy and NASA too, and it became the single arbiter of internetworking standards for the federal government.

With privatization, and with the spread of the Internet around the world, it became politically necessary to move the system's technical administration out of the US government. In January of 1992, the Internet Society, a nonprofit organization, was assigned formal oversight of the IAB and the IETF. In addition, the Internet Society took on the task of disseminating information about the Internet to the general public. The Internet Society, the IAB, and the IETF included members from all sectors of the Internet community, and international participation in these three groups increased over the course of the 1990s.

Throughout these changes, the Internet's administrative and technical structures remained remarkably decentralized. No one authority controlled the operation of the entire Internet. Drawing on the examples provided by the ARPANET culture and by contemporary experiments with privatization, the Internet community evolved several principles for reducing chaos and conflicts of interest in a decentralized and heterogeneous system. These included having multiple competing service providers wherever feasible; designing the system to maximize the number of operational decisions that could be made at the local level; and, in cases such as protocol standards where it is necessary to have a single decision-making group, having that group be inclusive and democratic.

Yet it has continued to be difficult for the Internet community to work out management policies that satisfy every interest group. The administrative structures of the Internet have been in a state of flux ever since privatization, as different solutions have been tried, and the ultimate source of the authority of the Internet Society, the IAB, and the IETF remains uncertain. With unusual candor, a 1997 FCC policy paper noted the following:

The legal authority of any of these bodies is unclear. Most of the underlying architecture of the Internet was developed under the auspices, directly or indirectly, of the United States government. The government has not, however, defined whether it retains authority over Internet management functions, or whether these responsibilities have been delegated to the private sector. The degree to which any existing body can lay claim to representing "the Internet community" is also unclear. (Werbach 1997)

As the Internet becomes more of an international resource, the continued authority of the United States in administrative matters will, no doubt, be challenged more and more.

The Global Picture

Today, few if any countries are without at least one connection to the Internet. How did this worldwide expansion occur? Though the

Internet originated in the United States, it did not simply spread from the United States to the rest of the world. Rather, its global reach resulted from the convergence of many streams of network development. Starting in the 1970s, many other nations built large data networks, which were shaped by their local cultures and which often served as agents and symbols of economic development and national sovereignty. The question was not whether these countries would adopt an "American" technology; it was whether and how they would connect their existing national or private networks to the Internet.

Since the early 1970s the ARPANET and the Internet had included sites outside the United States; University College London had an ARPANET connection for research purposes, and ARPA's Satellite Network linked the United States with a seismic monitoring center in Norway. The defense portion of the Internet also connected many overseas military bases. But the Internet's ownership by the US government was an obstacle to connecting it with civilian networks in other nations. ARPA and NSF managers feared that such connections would be perceived by the American public as giving away a taxpayer-subsidized resource to foreigners, and citizens of other countries might regard the encroachment of US networks as a form of imperialism. Overseas, grassroots user-supported networks with lower political profiles, such as BITNET and UUCP, spread faster than the Internet.

Before privatization, therefore, it was difficult to expand the Internet abroad by adding host sites to the US-run networks; connecting the Internet to networks in other countries was much more promising. By the mid 1970s, state-run networks were being built in a number of countries, including Canada, Germany, Norway, Sweden, Australia, New Zealand, and Japan (Carpenter et al. 1987). In addition to these national networks, there were several efforts to build multinational networks across Europe in support of the creation of a European Union. These included the European Informatics Network (established in 1971) and its successor, Euronet. Some of the networks were, like the ARPANET, designed for research and education; others provided commercial network services.

France Telecom, with its Minitel system (introduced in 1982), was the first phone company to offer a network service that provided content as well as communications. Since few people in France owned or had access to computers at that time, the phone company encouraged widespread use of Minitel by giving its customers inexpensive special-purpose terminals they could use to access the system. Minitel allowed millions of ordinary people to access online telephone directories and other commercial and recreational services (including online pornography, a popular attraction that received much public comment and that the US-government-run Internet could not have openly supported).

One of the world's leading sites for computer networking was CERN, the European laboratory for particle physics. Owing to the peculiar needs of its users, CERN had a long history of networking (Carpenter et al. 1987). Experimentalists in high-energy physics must travel to accelerator sites such as CERN. While there, they generate huge amounts of data. In the early 1980s, to make it easier to transfer such data around its laboratory in Geneva, CERN installed local-area networks. Physicists also need to communicate with and transfer data to their home institutions. To accommodate this need, CERN joined various wide-area networks, including EARN (the European branch of BITNET), the Swiss public data network, and HEPNET (a US-based network for high-energy physics).

Networks outside the United States had few links to the Internet while it was under military control. But when the National Science Foundation set up its civilian NSFNET, foreign networks were able to establish connections to it, and thus to gain access to the rest of the Internet. Canada and France had connected their networks to the NSFNET by mid 1988. They were followed by Denmark, Finland, Iceland, Norway, and Sweden later in 1988; by Australia, Germany, Israel, Italy, Japan, Mexico, the Netherlands, New Zealand, Puerto Rico, and the United Kingdom in 1989; and by Argentina, Austria, Belgium, Brazil, Chile, Greece, India, Ireland, South Korea, Spain, and Switzerland in 1990 (MERIT 1995). By January of 1990 there were 250 non-US networks attached to the NSFNET, more than 20 percent of the total number of networks. By April of 1995, when the NSF ceased operating it, the Internet included 22,000 foreign networks—more than 40 percent of the total number (ibid., file history.netcount). The system had truly become international in scope, though its membership remained highly biased toward wealthy developed countries.

The other industrialized nations approached networking rather differently than the United States. In the United States, the federal government operated military and research networks, but public network services were provided on a commercial basis. In other countries, the public networks were government-run monopolies, so network decisions involved overtly political maneuvers as well as business considerations. In many countries, people viewed the expansion of US networks such as the Internet with alarm, seeing it as further evidence of US economic dominance in the computing industry. Thus, while many people inside and outside the United States favored expanding the Internet around the world, politically charged differences between network systems presented a number of barriers.

One technical obstacle was incompatibilities among network systems. Initially, many networks outside the United States had used proprietary network systems or protocols designed by their creators. Most state-run networks eventually adopted the official CCITT or ISO protocols, which they regarded as the only legitimate standards; few if any used TCP/IP.

In the mid 1980s, however, many private network builders outside the United States began adopting TCP/IP, perhaps because they had become impatient with the slow introduction of ISO standards. In November of 1989, a group of TCP/IP network operators in Europe formed RIPE (Réseaux IP Européens, meaning European IP Networks). Similar in concept to the CIX (and perhaps providing a model for that system), RIPE connected its member networks to form a pan-European Internet, each network agreeing to accept traffic from the other members without charge. RIPE also provided a forum for members to meet, discuss common issues, and work on technical improvements. By 1996, RIPE had as members more than 400 organizations, serving an estimated 4 million host computers (RIPE 1997).

While the Internet protocols were gaining popularity outside the United States, many network operators wanted to reduce the United States' dominance over the Internet. One contentious issue was the structure of the Domain Name System. Since the ultimate authority to assign host names rests with the administrators of the top-level domains, other countries wanted to have their own top-level domains. Responding to these concerns, ISO promoted a domain name system in which each country would have its own top-level domain, indicated by a two-letter code such as "fr" for France or "us" for the United States.[21] Within these top-level domains, national governments could assign lower-level domains as they saw fit. The new system provided autonomy and symbolic equality to all nations. However, the old

Internet domain names based on type of organization (educational, military, etc.) were not abolished. In the United States, most organizations continued to use them, rather than adopting the new "us" domain (Krol 1992, p. 28).

Since the Internet originated in the United States, its "native language" is English—a fact that has caused some resentment among other linguistic groups. The dominance of English on the Internet has led to political disputes over what is often seen as American cultural or linguistic imperialism. In the mid 1990s, for example, the French government, which had put in place a number of measures to maintain French-language content in the media, required every Web site based in France to provide a French version of its text. Internet users whose native languages do not use the Roman alphabet have struggled to get support for extended character sets (Shapard 1995).

Finally, the expansion of the Internet has been limited by global disparities in the telecommunications infrastructure that underlies network access. In 1991, for instance, the number of telephone lines per 100 inhabitants in industrialized nations ranged from 20 (in Portugal) to 67 (in Sweden); in much of South America, Africa, and the Middle East, there were fewer than 10 lines per 100 inhabitants, and China, Pakistan, India, Indonesia, and Tanzania—countries with a huge percentage of the world's population—had fewer than one line per 100 people (Kellerman 1993, p. 132). Clearly, the unequal distribution of wealth among nations will continue to shape the Internet's worldwide role. The Internet, as a medium of instantaneous communication, might overcome geographic distance, but it cannot simply erase political or social differences.

The World Wide Web

In the 1980s the Internet's infrastructure grew impressively, but network applications lagged behind: email and file transfer were still the most common activities, and there were few user-friendly applications to attract novices. One factor that discouraged wider use of the Internet was its drab text-only interface, which contrasted sharply with the attractive graphical interfaces found on many personal computers. CompuServe, America Online, and Prodigy took advantage of the personal computer's graphic capabilities to provide attractive, user-friendly interfaces, thus setting a precedent for providing online information that incorporated images. Some software developers were also trying to create more graphics-oriented interfaces for Unix workstations (notably the X Windows system, developed at MIT in the mid 1980s), but many users of time sharing machines were still confined to text-based network interfaces.

Another drawback to using the Internet was the difficulty of locating and retrieving online information. File-transfer programs were available, but the user had to know the names of the desired file and its host computer, and there was no automated way to get this information. In former times it had been the ARPANET Network Information Center's role to provide information on network resources, and even then the information it had provided had often been inadequate. The privatized Internet had no central authority to create a directory of resources, and in any case the size of the Internet would have made the task of maintaining such a directory impossible.

In the early 1990s, new services made it easier to locate documents on the Internet. One such service was the gopher system, developed at the University of Minnesota. The gopher software allowed information providers to organize their information in a hierarchy of related topics; users of the system could select topics from menus, rather than having to know and type in the names of computers and files. Another widely used system was the Wide-Area Information Server, developed by the Thinking Machines Corporation. Instead of using a menu system, WAIS allowed users to search for documents whose text contained specified words; the titles of the documents would be displayed, and the user could retrieve the documents (Schatz and Hardin 1994, pp. 895–896). Services such as gopher and WAIS took a step in the direction of organizing information by content rather than location. There were still many obstacles to finding information on the Internet, however. There was no way to link information found in different documents, and the various protocols that had evolved for exchanging information were not compatible; no one program could handle formats as diverse as ftp, mail, gopher, and WAIS.

All these issues were addressed by a new Internet application that became known as the World Wide Web. The Web would fundamentally change the Internet, not by expanding its infrastructure or underlying protocols, but by providing an application that would lure millions of new users. The Web also changed people's perception of the Internet: Instead of being seen a research tool or even a conduit for messages between people, the network took on new roles as an entertainment medium, a shop window, and a vehicle for presenting one's persona to the world.

Building the Web

The Web did not spring from the ARPA research community; it was the work of a new set of actors, including computer scientists at CERN, the staff of an NSF supercomputer center, and a new branch of the software industry that would devote itself to providing Web servers, browsers, and content.

The first incarnation of the Web was created in 1990 by Tim Berners-Lee, Robert Cailliau, and others at CERN. Berners-Lee appreciated the value of networking; however, he saw a severe limitation in the fact that, though personal computers were becoming increasingly image oriented, most uses of the Internet were still limited to text. He envisioned a system that would help scientists collaborate by making it easy to create and share multimedia data (Berners-Lee et al. 1994, p. 82; Comerford 1995, p. 71). CERN had adopted TCP/IP in the early 1980s in order to provide a common protocol for its various systems, so Berners-Lee designed the new service to run over the Internet protocols.

The computing tradition on which Berners-Lee drew was far removed from the military roots of the ARPANET and the Internet: the hacker counterculture of the 1960s and the 1970s. In 1974, Ted Nelson, a vocal champion of this counterculture, had written a manifesto, *Computer Lib*, in which he had urged ordinary people to learn to use computers rather than leaving them in the hands of the "computer priesthood." More to the point, Nelson had proposed a system of organizing information that he called "hypertext." Hypertext would make it possible to link pieces of information, rather than having to present the information in a linear way.

Berners-Lee planned to create a hypertext system that would link files located on computers around the world, forming a "world wide web" of information. To the idea of hypertext he added the use of multimedia: his system included not only text-based information but also images, and later versions would add audio and video. (See Hayes 1994, p. 416; Schatz and Hardin 1994.) The Web's use of hypertext and multimedia drastically changed the look and feel of using the Internet.

In Berners-Lee's vision, the Web would create "a pool of human knowledge" that would be easy to access (Berners-Lee et al. 1994, p. 76). Before achieving this goal, however, Berners-Lee and his collaborators had to address a number of technical challenges. First, they had to create a shared format for hypertext documents, which they called hypertext markup language (HTML).[22] To allow the Web to handle different data formats, the designers of HTML specified a process of "format negotiation" between computers to ensure that the machines agreed on which formats to use when exchanging information. "Our experience," Berners-Lee (1993a) observed, "is that any attempt to enforce a particular representation . . . leads to immediate war. . . . Format negotiation allows the web to distance itself from the technical and political battles of the data formats." Like the ARPANET's designers before them, the Web team chose to create a system that could accommodate diverse computer technologies.

The layered structure of the Internet meant that Berners-Lee could build his new application on top of the communications services provided by TCP/IP. His group designed the hypertext transfer protocol (HTTP) to guide the exchange of information between Web browsers and Web servers. To enable browsers and servers to locate information on the Web, there also had to be some uniform way to identify the information a user wanted to access. To address this need, they created the uniform resource locator (URL)—a standard address format that specifies both the type of application protocol being used and the address of the computer that has the desired data. An important feature of the URL was that it could refer to a variety of protocols, not just HTTP. This would make it possible to use the Web to access older Internet services, such as FTP, gopher, WAIS, and Usenet news. The accommodation of all Internet services—present and future—within a single interface would be an important factor in making the Web system versatile and user friendly (Berners-Lee et al. 1994, p. 76; Berners-Lee 1993b; Schatz and Hardin 1994, pp. 896–897).

In December of 1990 the first version of the Web software began operating within CERN. Berners-Lee's system was an instant hit with CERN users. It took more than an inspired invention, however, to create an application that would bring the Internet mass popularity. It also required the right environment: widespread access to the Internet (made possible by privatization) and the technical means for users to run the Web software (provided by the personal computer).

Personal computers had brought computing to masses of ordinary Americans in the 1980s, and a decade later they laid the foundation for the popular embrace of the Web. The popularization of the Internet could have occurred without the personal computer. France's widely used Minitel system, for instance, relied on inexpensive home terminals for its user interface. But Minitel did not allow users to create their own content—a distinctive feature of the World Wide Web. The Web depended on significant computer power at the user's end of the connection. In addition, the time and energy that individuals had invested in learning to use their personal computers would make it easier for them to acquire the skills needed to access the Web. Thanks to the spread of graphical user interfaces via the Macintosh and Windows operating systems, instructions such as "point and click" seemed obvious rather than perplexing to novice Web users. For non-expert users in particular, the Internet-based Web represented the convergence of personal computing and networking.

Participation Explodes

CERN began distributing its Web software over the Internet in the summer of 1991, and in 1992 several other high-energy physics sites set up Web servers (Berners-Lee et al. 1994, p. 76; Berners-Lee 1995; Cailliau 1995). Among these sites was the National Center for Supercomputing Applications at the University of Illinois, one of the original NSF supercomputer centers. The NCSA had been affected by an unexpected development in computing technology: the decline of the supercomputer. These large machines had seemed state-of-the-art in the early 1980s; by 1990, however, they had lost their appeal for many scientists, since microprocessor-based workstations could provide comparable computer power in a more convenient and much less expensive form. The NCSA found itself with personnel and resources but a dwindling sense of purpose. Marc Andreessen, then a member of the NCSA's computing staff, later commented: "Because it was a Federal undertaking, the supercomputing program had a life of its own. . . . NCSA was really trying to figure out what it was going to do and what its role would be." (Andreessen 1995) The center's staff, which had been involved in designing the original NSFNET, decided to increase its emphasis on networking. Network services represented a growing market, and Andreessen and others saw great potential in the Web, which at that time was being used only by a small group of researchers.

In 1993 an NCSA team led by Andreessen began developing an improved Web browser called Mosaic. Mosaic was the first system to include color images as part of the Web page, and these images could, like text words, be used as links (Schatz and Hardin 1994, p. 897). More important, Mosaic was available to a much larger group of users than the CERN web browser had been. Mosaic was designed to run on most workstations and personal computers, and it was distributed by the NCSA over the Internet free of charge.

When the NCSA officially released Mosaic to the public, in November of 1993, more than 40,000 copies were downloaded in the first month; by the spring of 1994 a million or more copies were estimated to be in use (Schatz and Hardin 1994, pp. 897 and 900). Once Mosaic was available, the system spread at a phenomenal rate. In April of 1993 there had been 62 Web servers; by May of 1994 there were 1248 (Berners-Lee et al. 1994, p. 80). In 1994 Andreessen and his team left the NCSA to work on a commercial version of Mosaic called Netscape. Netscape simplified the browser's user interface, increased its speed, and added security measures to support financial transactions (Smith 1995, pp. 198–200; Andreessen 1995). The many new features made Netscape browsers (which were also distributed free of charge) a more popular choice for users than the older Mosaic.

Once the Web became popular, other companies began offering commercial browsers, and new businesses sprang up offering services that made it easier to locate information on the Web. The original system had had no way to search the Web; users could only type in a URL or follow links from page to page. Hypertext links continued to provide one important way to find information on related topics, but new programs called "search engines" made it possible to search for topics, organizations, or people on the Web. This went far toward solving the long-standing problem of identifying resources on the Internet, and it gave users more control over the way information on the Web would be presented to them.

The Web completed the Internet's transformation from a research tool to a popular medium by providing an application attractive enough to draw the masses of potential Internet users into active participation. It solidified the Internet's traditions of decentralization, open architecture, and active user participation, putting in place a radically decentralized system of information sharing. On the Web, links between sites were made laterally instead of hierarchically, and each individual could be a producer as well as a consumer of information.

The appearance of personal-computer-based Web browsers coincided with the privatization of the Internet, providing an attractive application for the many users who suddenly had access to the network. Whereas the ARPANET's early users had been beset by difficulties in their attempts to use remote computers, laypersons encountering the Web for the first time found it relatively easy to master. The fact that users could themselves become publishers of Web-based information meant that the supply of Web pages increased along with the demand, further accelerating the growth of the system (Schatz and Hardin 1994, p. 901). The Web's exciting multimedia format and the seemingly endless stream of new features offered by entrepreneurial companies put the Web at the center of public attention in the late 1990s, by which time "the Internet" and "the Web" had become synonymous to many people.

The Legacy of a Protean Technology

If there is a constant in the history of the Internet, it is surprise. Again and again, events not foreseen by the system's creators have rapidly and radically changed how the network has been used and perceived. This protean nature—the ability to take on unexpected and unintended roles—has been largely responsible for the Internet's endurance and popularity, and it explains the network's best-known legacies: the introduction of packet switching and other new techniques and the establishment of a unique tradition of decentralized, user-directed development. Some historians have even seen the Internet as a fitting technological symbol of the "postmodern" culture of the late twentieth century, in which unified authorities give way to multiple stakeholders with complex and contradictory agendas.[23] If the Internet is a reflection of our times, it may be all the more valuable to know how this unusual system came to be and what has held it together.

This book has explored, in various ways, the protean character of the Internet. The story of the invention of packet switching illustrates how the same basic technique could be adapted to different circumstances, with very different results. The success of packet switching did not depend on the ability of Paul Baran, Donald Davies, or Lawrence Roberts to accurately foresee the future of networking; indeed, all three made many assumptions that turned out to be incorrect. Rather, it was Roberts's ability and decision to build a flexible, general-purpose system that allowed the ARPANET to become a large-scale, long-lived, and highly visible example of the "success" of packet switching.

In the late 1960s the field of computer networking was still in its infancy; there was little theory or experience to provide guidance, and computer and communications technologies were in the midst of rapid change. In building the ARPANET, Roberts and the other system designers managed this uncertainty by incorporating it as an element of the system. Rather than try to rationalize and neatly plan each aspect of the system, the ARPANET's builders designed it to accommodate complexity, uncertainty, error, and change. This was done both through technical choices (such as layering) and by making human beings, with their inherent adaptability, an integral part of the system.

The ARPANET's creators were able to answer the question of *how* to build a large computer network. They had a harder time demonstrating *why* people should use one. Users played a crucial part in making the ARPANET more than an elaborate experiment in packet switching. Applications created by users became parts of the infrastructure, thus eroding the boundary between user and producer. By adopting electronic mail rather than remote computing as their favored application, users created a system that met their own needs and provided a compelling argument for the value of networking.

The Internet program of the 1970s took the values of flexibility and accommodation of diversity even further. The Internet's TCP/IP protocols, gateways, and uniform address scheme were all designed to create a coherent system while making minimal demands on the participating networks. The hard work that so many ARPA contractors put into implementing TCP/IP in the early 1980s created a standard technology that could be used relatively effortlessly by those who came after. The Internet's builders were able to adapt to challenges from outside, too. Faced with competing international standards that promoted different networking paradigms, Internet supporters worked to incorporate rival standards such as X.25 within their own system and to change the international standards to more closely resemble the Internet model.

In the 1990s, the Internet proved adaptable enough to make the transition to private commercial operation and to survive the resulting fragmentation of authority. The Internet's decentralized architecture made it possible to divide operational control among a number of competing providers, while its open and informal structures for technical management were able (at least in the near term) to survive new commercial and political pressures. The astonishing success of the World Wide Web showed that the Internet remained a fertile ground for network innovations. The Web drew on new computing technologies (particularly the personal computer and its graphical user interface) and its promoters thrived in the new commercial environment for Internet services. The Web also continued the tradition of decentralized participation in the creation of the system, encouraging individual users to add new content and tools.

As to the future, the only certainty is that the Internet will encounter new technical and social challenges. If the Internet is to continue as an innovative means of collaboration, discovery, and social interaction, it will need to draw on its legacy of adaptability and participatory design.

Notes

Introduction

1. I use "managers" as a generic term for directors and program managers. All these individuals were directly involved in research management.

2. The first scholarly work to treat the history of the Internet, Arthur Norberg and Judy O'Neill's *Transforming Computer Technology* (1996), is a strong example of the institutional approach. In it, Norberg and O'Neill (see also their 1992 work) drew on specially arranged access to the ARPA archives and on an extensive oral-history project to present a detailed analysis of ARPA's forays into network research and development. That book is not, however, intended as a full-scale history of the Internet; its focus is on ARPA's role in managing research in computer science. Only one of four case studies is devoted to networks, and it does not follow networking developments beyond ARPA's involvement. In the years since the Internet became a media sensation, a number of more popular books have appeared that deal in some way with its origins, often in a heroic manner. Hafner and Lyon's journalistic history *Where Wizards Stay Up Late* (1996) focuses mainly on the early years of the ARPANET. For more technically detailed accounts, see Salus 1995 and Quarterman 1990.

3. A notable recent exception is Paul Edwards; see his 1996 book *The Closed World*.

Chapter 1

1. For examples of how computer scientists have linked the ARPANET with the development of packet switching, see Tanenbaum 1989 and Quarterman 1990.

2. For a discussion of "normal" technology, see Constant 1980.

3. For examples of accounts of the origins of packet switching, see Norberg and O'Neill 1996, pp. 161–166; Hafner and Lyon 1996, pp. 64, 76–77, 113; Abbate 1994.

4. In his first scene in the movie, Dr. Strangelove is asked about the feasibility of building a "Doomsday Machine." His reply begins: "I commissioned last year a study of this project by the Bland Corporation. . . . "

5. On p. 13 of the transcript of the 1990 interview Baran added this note: "The focus of all those that I knew [who] were concerned about the nation's defense was on avoidance of war. I never encountered anyone who deserved the Dr. Strangelove war monger image so often unfairly ascribed to the military." However, on page 1 of the introduction to his 1960 paper describing a survivable communications system Baran explicitly characterized his proposed network as a tool for recovering from—rather than forestalling—a nuclear war: "The cloud-of-doom attitude that nuclear war spells the end of the earth is slowly lifting from the minds of the many. . . . It follows that we should . . . do all those things necessary to permit the survivors of the holocaust to shuck their ashes and reconstruct the economy swiftly."

6. "Rand received its money once a year," Baran (1990, pp. 10–11) noted, "and it was allowed pretty much to do what it wanted to do." Individual researchers, Baran continued, could use that flexibility to pursue their own projects: "Rand was a most unusual institution in those days. If you were able to earn a level of credibility from your colleagues you could, to a major degree, work on what you wanted to. . . . Rand had what today would be considered a remarkable amount of freedom on how it would spend its money."

7. For a description of message switching in telegraphy and a discussion of its advantages over ordinary line switching, see *Data Processing* 1969. That message switching was still being presented as an innovation in 1969 suggests how much more radical packet switching would have seemed at that time.

8. Baran (1964a, volume V, section IV) acknowledged that his own system "requires a more complex routing logic at the nodes than has ever been previously attempted."

9. In fact, when Baran (1964a, volume V, section IV) compared his own method to other military message switching systems, he counted the fact that "it is fundamentally an all-digital system" as a "disadvantage," since an all-digital telephone network was not yet available. In addition to permitting more links per connection, digital transmission would have allowed Baran to use inexpensive microwave links in the network. Microwave transmission requires many repeaters between stations; in order to transmit an analog signal clearly, these repeaters would have to be engineered for low distortion, which would make the equipment fairly expensive. In a digital system, the signal could be regenerated at each station, permitting the use of cheaper repeaters (Baran 1964b, p. 5).

10. See also O'Neill 1985, pp. 531–532.

11. The term "packet switching" was coined by Donald Davies, but the word "packet" seems to have been in common use in the field of data communications. Baran himself used "packet" as a generic term for a small quantity of

data, but opted to use "message block" as his formal term (1964a, V, section IV): "All traffic is converted to digital signals and blocked into standardized-format packets of data. Each packet of data, or 'message block,' is rubber-stamped with all the signaling information needed by other stations to determine optimum routing of the message block."

12. The full-page abstract of Baran's work that appeared in *IEEE Spectrum* (August 1964) devoted most of its space to describing the redundant layout of the network; it did not even mention the advantages of packet switching.

13. In a letter to the author dated 18 June 1998, Baran points out that he realized that packet switching would also be useful for civilian communications and that as early as 1967 he proposed using high-speed networks for commercial applications, such as connecting consumers' TV sets to a "home shopping" service.

14. Edwards (1996, pp. 129–130) notes that the National Defense Act of 1947, which created the Office of the Secretary of Defense and the Joint Chiefs of Staff, was supposed to integrate the four services, but that central control was not aggressively pursued until Robert McNamara became Secretary of Defense in the 1960s. In the 1990s the Defense Communications Agency was renamed "Defense Information Systems Agency."

15. Of course, the Air Force might have objected to giving up control of the network regardless of the Defense Communications Agency's competence. Baran's system would have been difficult to build in any case, since he had designed it based on anticipated improvements in computing. When Baran made his proposal, the small, inexpensive computers he would have needed for his switching nodes did not even exist.

16. Baran (1990, pp. 25, 29) states that "those that were working in the community had early access as the work was developing" and that "in the community interested in communications survivability, there were no significant barriers to information flow."

17. This estimate was provided by Andy Goldstein of the IEEE History Center.

18. *The Economist* noted with approval that Wilson had exchanged "the old cloth cap for a vastly becoming new white coat" (quoted on p. 247 of Morgan 1992). For more reactions from the press, see p. 48 of Horner 1993. The idea of harnessing science to serve the public good had been a theme in Labour politics for some years, but it became a prominent issue only in the four or five years leading up to the 1963 election. For the 1959 election the party put out a pamphlet titled A New Deal for Science: A Labour Party Policy Statement. At its 1960 annual conference the party issued a new policy statement on the importance of science, which Wilson introduced with a speech that evoked the atomic bomb: "This is our message for the 60s—a Socialist inspired scientific and technological revolution releasing energy on an enormous scale and deployed not for the destruction of mankind but for enriching mankind

beyond our wildest dreams." (quoted on p. 57 of Horner 1993) The ruling Conservative Party also emphasized science, though on a more modest scale, in its 1963 Trend Report, which recommended more coordination of the government's R&D expenditures (Horner 1993, pp. 59, 66; Coopey 1993).

19. It is possible to have an interactive single-user machine (such as a personal computer) or a time sharing machine that does not serve users at interactive terminals.

20. "Dial-up" connections are the ordinary kind, where you dial the number you want to connect to. They are distinguished from leased-line telephone connections, where the connection is permanently set up (for the duration of the lease, anyway) and no dialing is needed.

21. Davies may have been even more aware of the cost of communications than his American colleagues. In the United Kingdom, in contrast with the United States, there was no flat rate for local telephone calls. Also, while American researchers tended to think in terms of academic computing (in which users normally accessed the machine from relatively short distances), the United Kingdom had a larger percentage of users who relied on distant commercial systems (Campbell-Kelly 1988, p. 225).

22. On the parallels between time sharing and packet switching, see p. 2 of Davies 1966b.

23. The Labour government specifically wanted to redirect R&D efforts away from military projects and toward civilian industry. In his 1963 Scarborough speech, Wilson had complained: "Until very recently over half of our trained scientists were engaged in defence projects or so-called defence projects." (quoted on p. 57 of Edgerton 1996)

24. A National Physical Laboratory memorandum discussing the use of typewriters as input devices for the network also invoked trade: "There is no English manufacturer of typewriters, so unless something is done about this we could find ourselves developing an elaborate computer network largely with the aid of American peripheral devices." (Wichmann 1967, p. 3)

25. Davies did not even mention redundancy as a consideration until he had read Baran, at which point he wrote: "It must be admitted that the redundancy features needed to ensure a service in the presence of faults have not been carefully thought out." (Davies 1966b, p. 21) A later NPL proposal (Wilkinson 1968) described a packet switching network with a more distributed layout.

26. This possibility is discussed on p. 3 of Davies 1966a.

27. Davies (1966b, p. 10) noted: "The complexity of the interface unit is a consequence of giving the communications system the job of assembling and distributing messages for many slow terminals."

28. Another member of the club commented: "The war with the GPO . . . would have been uproarious had it not been so serious." (Malik 1989, p. 52)

29. Davies later said this of the General Post Office (1986, p. 11): "I had been in contact with them enough to know they were a pretty large, monolithic organization, in which to get anything done you have to convince a lot of departments. The fact that my ideas had had any impression on them at all was to me rather amazing, and I didn't expect them to put a lot of effort into it quickly. What I thought was that by popularizing the idea outside, by showing them in the experiment that it would work, we would eventually be able to put pressure on the Post Office to do something, and that was how it turned out."

30. Two of the most important differences between EPSS and the NPL network were that EPSS required users to implement a complex protocol into order to link their computers to the nodes and that EPSS based its service on "virtual circuits" rather than on "datagrams" (Hadley and Sexton 1978, p. 101). The latter terms are explained in chapter 5 below.

31. Barber (1979) describes how the General Post Office ended up using foreign technology despite the Wilson government's pressure to "buy British."

32. The ARPANET was soon followed by a French packet switching network called Cyclades, which was also quite innovative. Cyclades is discussed further in chapter 4.

33. In 1972, ARPA was given the status of a separate agency within the Department of Defense, and its name was changed to "Defense Advanced Research Projects Agency." In 1993 the name was changed back to "Advanced Research Projects Agency," apparently to signal a renewed commitment to research that would benefit civilian as well as defense industries. For consistency, I use "ARPA" throughout this work except when quoting directly from sources where "DARPA" was used.

34. Flamm (1988) and Edwards (1996) have documented how computing developments in the US were largely driven by military considerations.

35. Davies later recalled: "We were extremely surprised when we saw the [ARPANET] design to discover that they used the same message packet size as us, and they had many things in common." (1986, p. 14) On other computer scientists who were influenced by Scantlebury and who advocated using the NPL techniques in the ARPANET, see p. 232 of Hafner and Lyon 1996.

36. In the autumn of 1973, another member of the NPL group visited BBN to discuss congestion-control mechanisms that had been developed for the NPL network (Bolt, Beranek and Newman 1973, p. 2).

37. Howard Frank, who designed the ARPANET topology, described his earlier work on network vulnerability as "the follow-on to the work that Paul

Baran did" (Frank 1990). Leonard Kleinrock, who had a contract for analyzing the ARPANET's behavior, said of Baran: "I was well aware of his results. In fact I quoted his results in my own [1963] dissertation." (Kleinrock 1990)

38. Robert Taylor has emphasized that surviving an attack was *not* the point behind the ARPANET's design (Hafner and Lyon 1996, p. 10). Some of the ARPANET spinoffs that were later built for the Department of Defense did try to incorporate the type of survivability and security features proposed by Baran (Heiden and Duffield 1982, pp. 61, 64, 67–68).

39. Baran (1964a, volume VIII, section VII) foresaw this: "The use of all-digital transmission and switching of standardized Message Blocks greatly facilitates the addition of new features thought desirable in future communications networks for military and civilian applications. The ease of providing these new services, in comparison to present-day practice, seems to make this new form of communication network desirable—even in those applications where no vulnerability problem exists."

Chapter 2

1. In a 1989 interview, Taylor recalled: "I became heartily subscribed to the Licklider vision of interactive computing. The 1960 "Man-Computer Symbiosis" paper had had a large impact on me. . . . I don't really know how many other people it influenced, but it certainly influenced me."

2. Years later, after the ARPANET had become widely acclaimed, Taylor (1989, pp. 43–44) confessed to having "blackmailed Larry Roberts into fame." See also a note on p. 145 of Roberts 1988.

3. Marvin Minsky and John McCarthy were well-known computer scientists of the time. Minsky, at MIT, did ARPA-funded research in artificial intelligence; McCarthy, at Stanford, had an ARPA contract to develop time sharing systems.

4. Richard G. Mills, director of MIT's information processing services, voiced a common attitude: "There is some question as to who should be served first, an unknown user or our local researchers." (Dickson 1968, p. 134)

5. See, e.g., Padlipsky 1983 and Crocker 1993. My analysis of ARPA's management strategies draws on the work of Norberg and O'Neill (1996), who document the history and management style of ARPA's Information Processing Techniques Office in much greater detail.

6. Programming languages provided one precedent for dividing a complex task into layers or "levels." The development of "high-level" languages such as FORTRAN, COBOL, and LISP in the late 1950s made it possible to write programs using words and symbols; to execute these programs, a computer had to first translate them into a "low-level" machine code that used binary numbers.

7. On the pros and cons of the IMP idea, see Roberts 1967a.

8. As we will see in chapter 5, the protocol stack model could also provide a basis for comparing alternative network systems.

9. When asked if ARPA's system was elitist in comparison with that of the National Science Foundation, the other major source of funding for computer science, Taylor (1989, pp. 26–27) replied: "I can just as easily argue that having a committee of folks make a decision about supporting a particular proposal that goes to the National Science Foundation is an example of elitism. . . . You float that proposal around to a group of people that the NSF has chosen to be its peer review group. And if you're in the club you might get accepted, and if you're not you might not get accepted. So, I think I reject the notion of elitism versus democracy, and just say there was a different set of objectives, a different modus operandi." Leonard Kleinrock, one of those fortunate enough to get ARPA funding, compared IPTO favorably with what he described as "the heavy effort in proposal writing, the long review process, the small amounts" involved in getting funds from the NSF (Kleinrock 1990, p. 40).

10. Defending the DoD's role in graduate education in a 1972 speech before Congress, the Director of Defense Research and Engineering, John Foster, argued that government-sponsored research projects were themselves educational: "Some have said that federal support of research in universities should be stopped because it takes the faculty away from the students. This misinterprets the process of graduate education in science, for in this area research and education are inseparable. . . . Most important, it introduces the graduate student to tough, real-world problems which he can perceive as worthy of the highest effort. It is through this process that we educate future scientists and engineers in the background and problems of our Nation, and thus assure a supply of knowledgeable people to tackle future problems." (Foster 1972, 354–355) As Foster's speech implies, part of the purpose behind funding graduate students was to get them interested in working on defense-related topics.

11. Howard Frank, who had worked for several other government agencies, felt that IPTO's informal style was feasible only because the agency was creating a system that was experimental rather than operational: "It's not like fielding an SST and trying to make 12 billion dollars' worth of equipment and people come together at a single point in time. You couldn't run a space program like that, for instance." (Frank 1990, p. 30)

12. Budget figures derived from US Congress 1968, p. 2348.

13. The NWG first met with a small number of people at SRI on 25 and 26 October 1968. See RFCs 3, 5, and 24 for some early membership lists. RFC 1000 lists the authors and topics of the first thousand RFCs.

14. In its reliance on interpersonal networks, the ARPANET was typical of contemporary patterns of military-industrial-academic interaction. Defense

research projects tended to be concentrated in a small number of institutions where personal connections played an important part. Half the money spent by all government agencies on university science in 1965 went to 20 institutions (Johnson 1972, p. 335). A 1965 study noted that the defense R&D industry was highly concentrated in New England and Los Angeles and found that in New England nearly two-thirds of the people working in defense R&D had gone to school in the same area; in Los Angeles the figure was 21 percent. Half of the engineers and scientists surveyed said they had sought their current job because they had a personal acquaintance in the company (Shapero, Howell, and Tombaugh 1965, pp. 30, 50–51).

15. The checksum technique takes advantage of the fact that computers represent all data numerically. The computer sending the message divides it into segments of a certain size, interprets the series of digits in each segment as a single number, and adds up the numbers for all the segments. This sum (the checksum) is then appended to the message. The receiver likewise divides the message up, adds up the segments, and compares the result with the number appended. If the two checksums are different, the receiver knows that the message has been corrupted.

16. Leonard Kleinrock, in his 1962 doctoral dissertation, was one of the first to suggest using a distributed routing algorithm.

17. IMPs were down 2 percent of the time on average, largely as a result of hardware problems; line failures caused a similar amount of network downtime (Ornstein et al. 1972, p. 252).

18. A 1972 report that drew on discussions with both AT&T and BBN personnel noted: "The degree of rapport between these groups with respect to restore activity is exceptional with relatively none of the finger pointing problems that are so common when independent elements are melded into one function. It is generally recognized that this environment is largely due to the proficiency of the network control center in correctly diagnosing problems and producing relatively few false alarms." (RCA Service Company 1972, p. 11)

19. Roberts also objected to the fact that the protocols were asymmetrical (one machine was treated as a "client" and the other as a "server"); he felt that a symmetrical relationship would be more general and flexible. According to Crocker, the NWG members, having "suffered [their] first direct experience with 'redirection' [from the ARPA management]," spent the next few months designing a symmetrical host-host protocol.

20. The telnet program makes a connection to a remote computer; it then displays the user's typed instructions and the computer's responses on the user's terminal. The main difficulty in developing telnet was the variety of terminals in use, ranging from simple teletypewriters to sophisticated graphics displays; it was impractical to try to equip telnet with the display instructions for every different type of terminal in the network. The NWG's solution was

to define a standard "virtual terminal"—originally known as a "universal hardware representation" (Rulifson 1969). In computing, the word 'virtual' is used to refer to a computer simulation of a physical condition. The virtual terminal was a model of the minimum set of display capabilities that most terminals were expected to have. Telnet would issue display instructions for the virtual terminal, and each host computer would translate these into commands specific to its particular terminals. The virtual terminal provided an intermediate step between the general functions performed by the telnet application and the specific commands that the host used to control its own hardware. By using a simplified abstraction to provide a common language for terminal commands, the virtual terminal scheme masked the complexity of hardware incompatibility. The file transfer protocol (ftp) used a similar approach: to avoid the need to translate between a number of different file formats, it used standard network file formats that could be recognized by all hosts.

21. One predicted failure mode was "reassembly lockup." When an IMP receives packets from the network, it must reassemble them into a complete message before it can pass them on to the host. The IMP has a limited amount of memory space in which to reassemble messages. Simulation showed that if this space filled up with half-completed messages the IMP would become deadlocked: in order to deliver any of these messages, it would have to receive and assemble all the packets for that message, but it would not be able to accept any more packets until it freed up space by delivering some of the messages. To avoid this, BBN revised the system to make sure that the IMPs reserved sufficient memory to reassemble long messages. Another potential problem, "store and forward lockup," was also detected through collaborative experiments (McQuillan et al. 1972, p. 742).

22. Source code is the high-level, human-readable form of a computer program. BBN wanted to distribute only the low-level binary or executable code, which would run on the IMPs but could not be understood or modified by people working on the network.

23. Frank (1990) went on to say that the BBN-NAC interaction "really was a gentlemanly kind of thing. . . . It was an adversarial relationship, but it was not a hostile adversarial relationship, if that is consistent."

24. In a 1990 interview, Kahn cited it as "the first attempt to bridge the gap among theory, simulation, and engineering" (Kahn 1990, p. 19).

25. In a conversation with author on 11 April 1997, John Day of BBN recalled that John Melvin of SRI, who was working on the project and trying to recruit students to participate, told them: "Look, our money's only bloody on one side!"

26. Lukasik (1973) also argued that the ARPANET gave small research groups a more equitable chance at getting DoD funding: "Before the network we were in many cases forced to select research groups at institutions with large

computing resources or which were large enough to justify their own dedicated computer system. Now it is feasible to contract with small talented groups anywhere in the country and supply them, rapidly and economically, with the computer resources they require via the network." (US Congress 1972, p. 822). Presumably this offered US Representatives from districts outside the country's main computing centers hope of bringing more defense research dollars into their districts.

27. One could argue that the need to be able to identify a military application does represent an "imposition," whether or not the researchers themselves recognized it as such.

28. In 1973 a data processing industry newsletter reported: "ARPANET has proven the technology—so private investors are lining up to back commercial versions." (McGovern 1973, p. 5)

29. In addition to publishing many individual articles about the ARPANET, computer journals and conference proceedings periodically highlighted ARPA's contributions to networking by featuring several ARPANET papers in a single issue. See especially *AFIPS Spring Joint Computer Conference,* 1970 and 1972; *AFIPS National Computer Conference,* 1975; and *Proceedings of the IEEE* 66, no. 11 (1978).

30. See Kleinrock 1976. On Kleinrock's influence, see Frank, Kahn, and Kleinrock 1972, p. 265; Tanenbaum 1989, p. 631.

Chapter 3

1. For example, see the chapters by Cowan, Pinch and Bijker, Callon, and Law in Bijker, Hughes, and Pinch 1987.

2. The term "cyberspace" was coined by William Gibson in his 1984 novel *Neuromancer* (Ace Books). Most current uses of the term have little or no relation to Gibson's ideas. Gibson was not the first to envision a virtual world based on computer networks; for an earlier story that anticipates many current issues, see Vernor Vinge's 1979 novella "True Names," in Vinge, *True Names and Other Dangers* (Baen Books, 1987).

3. This cost included $45,000 for an IMP or $92,000 for a TIP plus $10,000–$15,000 for the hardware interface between the IMP or TIP and the site's host machine.

4. Much of this section is based on conversations with Alex McKenzie, who ran BBN's Network Control Center, and on a 1972 report by the RCA Service Company, which ARPA had commissioned to assess the ARPANET's status and make recommendations for improvements.

5. For an example of how the lack of site information was still a problem as late as 1981, see Haughney 1981a.

6. Successive generations of Internet Network Information Centers have continued to archive RFCs.

7. For example, when the Network Working Group was developing the telnet protocol, which allows users to log in to a remote computer, it took them months just to decide how to indicate the end of a line of input. Each operating system had its own conventions for handling input, and this made it hard to design a workable protocol without forcing some sites to alter the behavior of their systems. (Examples of disparities included full vs. half duplex echoing, use of "line feed" and "carriage return" characters, and whether or not the password would appear on the screen when the user typed it in.) If such a simple action as ending a line of input was so difficult to standardize, providing a general-purpose network interface for complex input/output devices would clearly be even harder (McKenzie 1997).

8. 'Elf' is German for eleven, as in PDP-11; the name is also a play on IMP.

9. On the popularity of the "utility" model, see p. 416 of Massy 1974. McKenzie (1990, p. 13) compared the network's operation to a utility such as a power company. Similarly, in a 1974 article, the vice-president of DEC, C. Gordon Bell, argued that "the next logical step" after the success of the ARPANET would be "an information utility for whole 'communities,' such as businesses, homes, and government departments" (Bell 1974, p. 44).

10. MIT's Information Processing Services, which managed the MULTICS machine, apparently was worried that local users would end up subsidizing ARPANET users. This financial and ideological concern had adverse technical effects on the system. MIT's system managers felt that if the ARPANET protocols ran at the operating system level they would add to the general system "overhead" that was paid for by all users; on the other hand, protocols that ran at the user level could be billed to individual users. Therefore, they tried to implement the protocols in a way that would confine most of the network activity to the user level, even though this interfered with the efficient operation of the protocols (Heart et al. 1977, III-24–III-25).

11. The use of the ARPANET to develop new networking techniques is discussed in more detail in chapter 4 below.

12. The main focus of the program was the short-lived Nile Blue project. Apparently, some in the US military believed that the USSR might be able to alter weather patterns in the US in disastrous ways—for instance, by causing droughts or storms. The purpose of the Nile Blue project was to find out whether such deliberate climate modification was possible and, if so, whether there could be any defense against it. After two or three years, it became apparent that the climate models were not good enough to settle this question, and that, even if the models worked, the computers of the time would not be able to process the vast amounts of data required. The project was discontinued, but the organizations that had been involved in it had gained valuable experience in climate modeling and had developed a number of useful tools for analyzing and displaying meteorological data.

13. For more on the seismic program (code named VELA), see Kerr 1985.

14. One attempt to set up such a system was the Resource Sharing Executive (RSEXEC), which BBN created for its TENEX operating system (Heart et al. 1977, p. III-76).

15. The RCA Service Company's 1972 report on the ARPANET made a similar point (p. A-82): "Sophisticated utilization of multiple computers for a single application . . . would depend on having the individual services at a reliable and accessible enough level to reasonably permit someone to come in and choose his tools. . . . From a technical point of view that threshold has been reached, however much remains to be done from a documentation and management point of view."

16. Frank Heart judged the ability to send messages to many users at once "perhaps the most important factor in the use of mail," because it facilitated group communication (Heart et al. 1977, p. III-670). Equally important for some users was the asynchronous nature of email. For ARPA's seismologists, the time zone differences among the stations in Montana, Virginia, and Norway made email especially attractive (Lukasik, telephone conversation with author, 1 May 1997; Dorin and Eastlake 1976).

17. McKenzie recalls this occurring around 1973 or 1974.

18. See also Licklider and Vezza 1978, p. 1331.

19. Les Earnest (email to author, 28 March 1997) noted: "I was surprised at the way the use of email took off, but so were the others who helped initiate that development. . . . We thought of [the ARPANET] as a system for resource sharing and expected that remote login and file transfer would be the primary uses."

20. Alex McKenzie pointed out to me the importance of mailing lists in building "virtual communities" on the ARPANET.

Chapter 4

1. 'Architecture' refers to the overall structure of the system and the relations between its parts.

2. In 1979, Kahn would become IPTO's director.

3. No routing is required in the simple case. If a broadcast system covers a large area or includes many sites, it may be subdivided into regions, in which case a routing system *is* needed to get packets from one region to another.

4. In a play on IMP, the interface was named after "a legendary Hawaiian elf" (Abramson 1970, p. 282).

5. The project was coordinated by the Linkabit Corporation in San Diego; BBN provided the satellite IMPs and a Satellite Monitoring and Con-

trol Center; and UCLA analyzed network data. The Lincoln Lab was also involved.

6. "Eventually," Kahn (1990) recalled, "we all took the Internet technology pieces and created a separate program in DARPA for it. But originally, all that work was done as part of the packet radio program."

7. "Internet" was not adopted as the standard term for a set of connected networks until the early 1980s. Before then a variety of terms were used, including "virtual network," "multinetwork environment," and "catenet" (short for concatenated network).

8. Cerf left ARPA late in 1982 for a position at MCI, leaving Kahn once again in charge of the Internet Program.

9. The INWG also worked out specifications for its own proposed host protocol standard, but the group was unable to interest the international standards bodies in this proposed protocol (Cerf et al. 1976, p. 63).

10. McKenzie (1997) pointed out this difference between the BBN approach and the Internet design. In virtually all developed countries other than the US, state-owned Post, Telephone and Telegraph monopolies were planning to build packet switching networks in the mid 1970s. The PTTs modeled these networks on the voice telephone system, which had a sophisticated and reliable network and only limited functionality at the endpoints—just the opposite of the Cyclades model. This situation helps explain why the French network researchers were more vociferous about the need for a simple network design than the Americans, who did not have to counter an opposing paradigm advanced by an established power. The conflicts that arose between the PTTs and the computer-oriented network builders will be addressed in chapter 5.

11. The main nodes of EIN were built in Italy, France, Switzerland, and the United Kingdom (Laws and Hathway 1978, pp. 275–276). EIN was also used to test the proposed host protocol developed by the INWG (Sunshine 1981, p. 69).

12. Pup was never successfully launched as a commercial product, but a later version called the Xerox Network System (XNS) entered the market in 1980 and was widely used. XNS, in turn, was the basis for the popular Novell NetWare system of the 1990s (Metcalfe 1996, p. xix).

13. Notably absent from this group were representatives of Bolt, Beranek and Newman, who might have taken the opposing view that the network should provide reliability (as BBN's packet switches had done for the ARPANET). Some ARPANET veterans have intimated that BBN was deliberately excluded from the Internet project because other members of the network community resented its efforts to control all aspects of the ARPANET's operation. It is certainly true that the decentralized control favored by the Internet's designers would make it difficult for any single network operator to dominate the system. Thus, a new network architecture may have been seen as a chance to re-negotiate the balance of control among ARPANET participants.

14. Metcalfe (1996, p. xix) mentions his involvement in this seminar. Cerf and Kahn (1974) acknowledge the input of Davies and Scantlebury of the NPL and Pouzin and Zimmerman of Cyclades in the design process.

15. The translation was done using a technique called "encapsulation" that may have originated in Xerox's Pup effort (Cerf 1980, p. 11).

16. Strictly speaking, ARPANET packets were sent to a particular port on a particular IMP, since an IMP might have more than one host attached.

17. In theory, member networks only had to run IP; TCP was optional, though most host computers did use it.

18. Unix workstations were widely used at universities and research laboratories; thus, adding TCP/IP to Unix greatly expanded the availability of the Internet protocols and helped them become a de facto standard in academic networking (McKenzie 1997). The widely distributed version of Unix created at UC Berkeley is sometimes referred to as "BSD [Berkeley Standard Distribution] Unix."

19. These were the Community Online Intelligence System ("COINS"), begun around 1972, and the Platform Network, built in the late 1970s. According to Eric Elsam, who managed network projects at BBN's Washington office, both systems provided regular data communications service for intelligence agencies for many years (Elsam, telephone conversation with author, 22 July 1997).

20. The Air Force had sponsored much research on command and control systems, including Paul Baran's work on packet switching at Rand. Charles Zraket, who headed the MITRE Corporation's Washington Office (which was responsible for building WWMCCS), believed that the Cuban Missile Crisis was responsible for increasing the military's interest in computer networks. "There was a tremendous interest then in the DoD to use computers and communications for command and control. [The crisis] really spread it out from the Air Force, which had been pioneering these developments, throughout the whole Defense Department. . . . And both the Project MAC and the DARPA network efforts really helped in that respect, because these were technologies that people were able to pick up all over." (Zraket 1990) For a description of the Strategic Air Command system that preceded WWMCCS, see p. 107 of Edwards 1996.

21. DCA personnel had initially been critical of Lawrence Roberts's plans to use packet switching in the ARPANET. Eric Elsam, who was BBN's project manager for WIN, believed that the success of WIN gave the DCA a more favorable view of packet switching. (Elsam, telephone conversation with author, 22 July 1997)

22. RCA noted in its November 1972 report that ARPA was planning to solicit bids to run the network commercially, but that this depended on ARPA's being able to work out an arrangement with the FCC (RCA Service Company 1972, pp. A-44 and A-81).

23. Vint Cerf, who was still at Stanford University, contributed to this report as a consultant.

24. George Heilmeier, who became director of ARPA in 1975, argued that the network offered a way for "the users, not the engineers" to evaluate computer systems "in real command and control scenarios while injecting the all-important human factors" (Heilmeier 1976, p. 6). Kahn (1989, p. 22) recalls: "We put the technology in place and worked with the military services to see what they would do with it, what the impact was on their capabilities and operation."

25. The Air Force Systems Command planned to use the ARPANET to connect computers at its Wright-Patterson, Eglin, and Kirtland bases, thereby forming what it called the AFSCNET. Noting that the ARPANET had been a research-oriented system, one member of the Air Force computing staff commented: " . . . With its transition from ARPA support, [the ARPANET] seems to be becoming more operationally oriented. The proposed AFSCNET is an example" (Lycos 1975, p. 177).

26. See, e.g., Haughney 1980b.

27. BBN made a bid to build AUTODIN II on the basis of the ARPANET design; however, the DCA, evidently not yet wholly committed to the AR-PANET techniques, awarded the contract to Western Union. The Western Union design did not have the ARPANET's distributed, redundant structure—it had only eight switching nodes whereas the ARPANET had 58 by that time—but it was nonetheless touted as "designed to a higher level of reliability, survivability and throughput than ARPANET" (Kuo 1978, p. 310).

28. Writing a correctly functioning version of the TCP software was the hardest task. The BBN staff had considerable experience in writing TCP software by 1980, but it still took a BBN programmer 18 months to implement the new standard (Sax 1991).

29. The old Honeywell computers that had been used for the IMPs were replaced by BBN C/30s.

30. Elizabeth (Jake) Feinler (1982a) mentions the Internet Working Group (especially Jon Postel at ISI), the NIC, BBN, and the DCA's engineering group as participants in this project.

31. Eric Elsam, BBN's program manager for the Defense Data Network, described reluctant military users as being "dragged kicking and screaming" into the DDN system by DCA managers (Elsam, telephone conversation with author, 22 July 1997). On the civilian side, one ARPANET host administrator commented: "It is likely that if DCA didn't do this [i.e., enforce the transition to TCP], that NCP would still be the standard protocol." (Crispin 1991)

Chapter 5

1. It was not until 1981, when IBM brought out a personal computer that was not based on proprietary standards, that other manufacturers could

legally make and sell "clones" of an IBM computer. "Third party" vendors promptly took over most of the PC market from IBM.

2. For example, in the early 1990s standards for high-definition TV became a site of economic competition among the US, Europe, and Japan. See Neil, McKnight, and Bailey 1995.

3. Gordon Bell of DEC recalled: "We saw the need for [packet switching] once we built all those minicomputers. We had to have DECNET . . . and our model was all of the packet switching that was done in the community. We did it a little bit differently . . . but the ARPANET was clearly the model." (Goldberg 1988, p. 170)

4. In the case of data communications standards, ANSI also consulted with the Institute for Electrical and Electronics Engineers (a professional organization) and the Electronic Industries Association (a trade group).

5. This body is now known as the National Institute for Standards and Technology.

6. Many of the standards bodies discussed here have since been reorganized and renamed. In the early 1990s, the CCITT became the International Telecommunications Union Committee on Telephony (ITU-T). For consistency, I will use "CCITT" throughout this discussion.

7. See also Lynch and Rose 1993, p. 11.

8. Neither side in the debate seemed to be in favor of combining the two sets of protocols. X.25 supporters claimed that, with the PTTs providing dependable virtual circuits, end-to-end error checking and re-transmission by a host-based transport protocol such as TCP would be redundant (Quarterman 1990, p. 191; Davies and Bates 1982, p. 21; Blackshaw and Cunningham 1980, pp. 420–421). TCP/IP supporters countered that, since the only way hosts could count on reliable service was to be prepared to compensate for potentially unreliable networks, there was no real point in having elaborate error-control procedures in the network protocol (Cerf and Kirstein 1978, p. 1403).

9. See also p. 245 of Padlipsky 1983. Two popular networking texts not written by OSI supporters but nevertheless organized around the OSI model are Tanenbaum 1989 and Quarterman 1990.

10. In fact, the DoD had initiated its own data communications standards effort in 1978, partly to ensure consistent standards throughout the military, but also to make sure that military requirements would be "considered by the pertinent standardization forums" and, conversely, that DoD standards would be "responsive to emerging federal, national, and international standards" (Haughney 1980c). Evidently this initial effort had not been enough.

11. One PTT representative suggested that protocols developed by a "special interest group" such as ARPA would automatically be rejected by competing groups, and would in addition be too "task-specific" for general use (Black-

shaw and Cunningham 1980, pp. 417–418). Some US authors have said that ISO feared that US manufacturers would unfairly benefit from the adoption of TCP, since they had already developed products that used the ARPA protocols (Lynch and Rose 1993, p. 11).

12. It was officially designated "A Subnetwork Independent Convergence Protocol" (ISO 1984b, p. 9).

13. In this sense, Internet gateways are an example of a "gateway technology," defined by David and Bunn (1988, p. 170) as "some means (a device, or a convention) for effectuating whatever technical connections between distinct production subsystems are required in order for them to be utilized in conjunction. . . . A gateway technology, therefore, achieves technical compatibility in order to affect linkage or communication among subsystems."

Chapter 6

1. That most schools chose PhoneNet even though it offered only minimal service suggests that cost was indeed a factor in determining access.

2. See chapter 4 above.

3. Geoff Goodfellow, manager of an ARPANET host system, was able to keep his own host table more up to date than the NIC's. Whereas the NIC had to go through formal procedures to add hosts to its table, Goodfellow simply checked for the presence of new hosts on the network. According to Goodfellow, other host administrators heard that his machine had the most current information on ARPANET hosts, and a large percentage of sites began using his host table in favor of the NIC's (Goodfellow 1997).

4. The domain idea itself had been discussed in the networking community since the late 1970s, and earlier RFCs had proposed possible ways to implement it in the Internet.

5. This description is somewhat oversimplified; in most implementations, the name server would cache frequently requested addresses for the sake of efficiency.

6. This was done by putting the name of the non-TCP/IP network in place of a top-level domain—e.g., "host.bitnet." When asked for the address of such a name, the name server would instead return the address of a mail gateway that provided an interface between the Internet and the specified network. The mail gateway would translate the Internet-style address into the address format for the other network.

7. This has also had the effect that host names convey some information about the site. Conversely, a person who already knows something about an organization can often guess its domain name (e.g., microsoft.com or stanford.edu)—something that is not possible with zip codes or area codes. By choosing

meaningful rather than arbitrary designations for domains, the Internet's designers increased its usability.

8. The information on NSF regional networks is from pp. 301–338 of Quarterman 1990.

9. See Wolff 1991. The granting of contracts to IBM and MCI was somewhat controversial within the Internet community; Wolff notes that there was "widespread skepticism" about this award, since none of the companies involved had any TCP/IP experience.

10. By 1988 the Internet was estimated to include more than 400 networks, up to 500,000 hosts, and perhaps a million users around the world (Quarterman 1990, p. 278).

11. The 1992 version of the NSF's Acceptable Use Policy is reprinted on pp. 353–354 of Krol 1992.

12. This was a long-standing issue. Years earlier, the RCA Service Company (1972, p. A-83) had noted that political and legal obstacles prevented the ARPANET from being connected to other research networks, such as Michigan's MERIT network or the IBM-sponsored TSS Network: "Connection of the MERIT system would be looked upon by Congress as providing subsidies to a state resource. The legal ramifications of TSS connection to the network present nearly insurmountable obstacles."

13. Another factor that favored competition was that the NSF wanted to take advantage of recent technical advances in data communications, such as frame relay and asynchronous transfer mode switching, and they felt that having several backbone providers would encourage the use of a greater variety of new techniques.

14. PSINet's corporate web page gives an interesting glimpse of how the US business community viewed the rather anarchic structure of the Internet—which has been much celebrated by academic users—with skepticism: "Few in the mainstream corporate world in 1989 knew much about the Internet, and fewer still viewed it as a potential part of their own IT solutions. After all, the roots of the Internet were in the war room and the classroom, not in the boardroom. Worse, no one owned the Internet. No one controlled or managed it. No one was responsible for its performance. Why would any organization . . . entrust the delivery of their information to such a technology?" (PSINet 1997) To some extent this is a projection of present conditions on the past, since in 1989 the NSF still "owned" and managed the Internet (or, at least, its backbone).

15. The NSF actually split network operations into two categories: supplying the backbone infrastructure and overseeing the routing system. The latter task involved maintaining a database of domain name servers that was used by routers throughout the Internet to locate hosts. The NSF decided that the routing authority, which was both technically demanding and critical to the

stability of the system, should remain within a single organization, whereas the provision of backbone services could be divided among several bidders. MERIT eventually received a five-year award to continue its role as Routing Authority (Wolff 1991, p. 4).

16. In a conversation with the author in December of 1997, Robert Morris pointed out that NSF managers probably looked to the telephone industry for inspiration on how to design a system with multiple competing service providers. The US telephone system had been drastically reshaped in the early 1980s by the breakup of the AT&T monopoly. In the new deregulated system, AT&T and other long-distance carriers were supposed to connect on an equal basis with the local telephone networks, now run by several Bell Operating Companies. The telephone network became, in essence, an "internet" of local and long-distance telephone networks, with competing long-distance "backbones." This was exactly the structure that the NSF was trying to create, and it would have provided an obvious model for the redesign of the Internet—especially since MCI, Sprint, and other phone companies were involved in providing Internet services.

17. These gateway operators included PacBell in San Francisco, Ameritech in Chicago, Sprint in New York, and Metropolitan Fiber Systems in Washington, DC (MERIT 1995).

18. By 1982, mail relays had been set up between ARPANET and the commercial Telemail service, and between ARPANET and the British system NIMAIL (Postel, Sunshine, and Cohen 1982, p. 978).

19. Source of examples: Quarterman 1990, pp. 257, 233. In the first example, the user is sending a message from FidoNet to BITNET; in the second, from BITNET to JUNET.

20. The chairman of the IAB held the title "Internet Architect."

21. According to Barry Leiner (email to author, 29 June 1998), the country-code system had actually been envisioned by the original designers of DNS, but the impetus to adopt it as the standard way of designating domains seems to have come from outside the United States.

22. HTML was an based on an existing ISO standard called the Standard Generalized Markup Language. SGML is specified in ISO Standard 8879 (1986); HTML is specified in RFC 1866 (1995).

23. The Internet can thus be seen as an example of a "postmodern" technological system—i.e., one in which the unified operating authority is replaced by a decentralized, contradictory, and even chaotic form of control. See Hughes 1998 for a discussion of postmodern technological systems.

Bibliography

The library of Bolt, Beranek and Newman (now GTE Internetworking) is located at the company's main office in Cambridge, Massachusetts.

The Charles Babbage Institute is located at the University of Minnesota in Minneapolis. The archives are open to all researchers, and many materials are available online. For information, see http://www.cbi.umn.edu.

The National Archive for the History of Computing is part of the Centre for the History of Science, Technology & Medicine at the University of Manchester, UK. The archive is open to researchers. For information, see http://www.man.ac.uk/Science_Engineering/CHSTM/nahc.htm.

The sources designated RFC (Request for Comments) are electronic documents that define or discuss network standards and operating procedures. They are maintained at one or more Network Information Centers. For an online archive of RFCs, see http://www.rfc-editor.org.

Abbate, Janet. 1994. From ARPANET to Internet: A History of ARPA-Sponsored Computer Networks, 1966–1988. Ph.D. thesis, University of Pennsylvania.

Abramson, Norman. 1970. "The ALOHA System—Another Alternative for Computer Communications." In Proceedings of AFIPS Fall Joint Computer Conference. AFIPS Press.

ACM Computer Communication Review. 1975. "US Government Communications Network Activities." *ACM Computer Communication Review* 5 (4): 32–33.

Agre, Philip E. 1998a. "The Internet and public discourse." *First Monday* 3 (3): 00–00.

Agre, Philip E. 1998b. "The architecture of identity: Embedding privacy in market institutions." In Proceedings of the Telecommunications Policy Research Conference, Alexandria, Virginia.

Andreessen, Marc. 1995. Interview by David K. Allison, Mountain View, California, June. National Museum of American History, Smithsonian Institution.

Anonymous (probably K. A. Bartlett). 1967. NPL Communications System: Facilities Offered by Low and Medium Speed Terminal Hardware. Memorandum, National Physical Laboratory. In National Archive for the History of Computing.

Anonymous (probably Robert Kahn, Robert Metcalfe, Abhay Bhushan, and others). 1972. Scenarios for Using the ARPANET at the ICCC. McKenzie box 1, Bolt, Beranek and Newman library.

Aspray, William, and Bernard O. Williams. 1994. "Arming American Scientists: NSF and the Provision of Scientific Computing Facilities for Universities, 1950–1973." *Annals of the History of Computing* 16 (4): 60–74.

Aufenkamp, D. D., and E. C. Weiss. 1972. "NSF Activities Related to a National Science Computer Network." In *Proceedings of International Conference on Computer Communication.* North-Holland.

Baran, Paul. 1960. *Reliable Digital Communications Systems Using Unreliable Network Repeater Nodes.* Report P-1995, Rand Corporation.

Baran, Paul. 1964a. *On Distributed Communications.* Twelve volumes. Rand Report Series.

Baran, Paul. 1964b. "On Distributed Communications Networks." *IEEE Transactions on Communications* 12: 1–9.

Baran, Paul. 1990. Interview by Judy O'Neill, Menlo Park, California, 5 March. Charles Babbage Institute.

Barber, D. L. A. 1969. Visit to Canada and the USA, 22 September–17 October. Memorandum, National Physical Laboratory. In National Archive for the History of Computing.

Barber, D. L. A. 1979. A Small Step for Britain; A Giant Step for the Rest? Memorandum, National Physical Laboratory, April. In National Archive for the History of Computing.

Bell, C. Gordon. 1974. "More Power by Networking." *IEEE Spectrum,* February: 40–45.

Bell, C. Gordon. 1988. "Toward a History of (Personal) Workstations." In *A History of Personal Workstations,* ed. A. Goldberg. ACM Press.

Bennett, Christopher J., and Andrew J. Hinchley. 1978. "Measurements of the Transmission Control Protocol." *Computer Networks* 2: 396–408.

Berners-Lee, Tim. 1993a. "W3 Concepts." Web page http://www.w3.org/pub/WWW/Talks/General/Concepts.html.

Berners-Lee, Tim. 1993b. "W3 Protocols." Web page http://www.w3.org/pub/WWW/Talks/General/Protocols.html.

Berners-Lee, Tim. 1995. "Tim Berners-Lee." Web page http://www.w3.org/pub/WWW/People/Berners-Lee-Bio.html/Longer.html.

Berners-Lee, Tim, Robert Cailliau, Ari Luotonen, Henrik Frystyk Nielsen, and Arthur Secret. 1994. "The World-Wide Web." *Communications of the ACM* 37 (8): 76–82.

Bijker, Wiebe E., Thomas P. Hughes, and Trevor Pinch, eds. 1987. *The Social Construction of Technological Systems.* MIT Press.

Binder, R., N. Abramson, F. Kuo, A. Okinaka, and D. Wax. 1975. "ALOHA Packet Broadcasting—A Retrospect." In *Proceedings of AFIPS National Computer Conference.* AFIPS Press.

Blackshaw, R. E., and I. M. Cunningham. 1980. "Evolution of Open Systems Interconnection." In *Proceedings of Fifth International Conference on Computer Communication.* North-Holland.

Blanc, Robert P. 1986. "NBS Program in Open Systems Interconnection (OSI)." *Lecture Notes in Computer Science* 248: 27–37.

Blanc, R. P., and J. F. Heafner. 1980. "The NBS Program in Computer Network Protocol Standards." In *Proceedings of the Fifth International Conference on Computer Communication.* North-Holland.

Boggs, David R., John F. Shoch, Edward A. Taft, and Robert M. Metcalfe. 1979. Pup: An Internetwork Architecture. Report CSL-79–10, Xerox Palo Alto Research Center.

Bolt, Beranek and Newman. 1973. Interface Message Processors for the ARPA Computer Network. Report No. 2717.

Bouknight, W. J., G. R. Grossman, and D. M. Grothe. 1973. "The ARPA Network Terminal System: A New Approach to Network Access." In *Proceedings of IEEE DataComm 73.*

Brinton, James. 1971. "ARPA Registers a Big Net Gain." *Electronics,* 20 December: 64–65.

Broad, William J. 1983. "Global Computer Network Split as Safeguard." *New York Times,* 5 October.

Brock, Gerald. 1975. "Competition, Standards and Self-Regulation in the Computer Industry." In *Regulating the Product,* ed. R. Caves and M. Roberts. Ballinger.

Burdick, E., and H. Wheeler. 1962. *Fail-Safe.* McGraw-Hill.

Cailliau, Robert. 1995. "A Little History." Web page http://www.cern.ch/ CERN/WorldWideWeb/RCTalk/history.html.

Callon, Ross. 1983. "Internetwork Protocol." *Proceedings of the IEEE* 71 (12): 1388–1393.

Campbell-Kelly, Martin. 1988. "Data Communications at the National Physical Laboratory (1965–1975)." *Annals of the History of Computing* 9 (3/4): 221–247.

Campbell-Kelly, Martin, and William Aspray. 1996. *Computer: A History of the Information Machine*. Basic Books.

Carlyle, Ralph Emmett. 1988. "Open Systems: What Price Freedom?" *Datamation*, 1 June: 54–60.

Carpenter, B. E., F. Fluckiger, J. M. Gerard, D. Lord, and B. Segal. 1987. "Two Years of Real Progress in European HEP Networking: A CERN Perspective." *Computer Physics Communications* 45: 83–92.

Carr, C. Stephen, Stephen D. Crocker, and Vinton G. Cerf. 1970. "Host-Host Communication Protocol in the ARPA Network." In *Proceedings of AFIPS Spring Joint Computer Conference*. AFIPS Press. 589–597.

Cerf, Vinton G. 1979. "DARPA Activities in Packet Network Interconnection." In *Interlinking of Computer Networks*, ed. K. Beauchamp. Reidel.

Cerf, Vinton G. 1980. "Protocols for Interconnected Packet Networks." *Computer Communication Review (ACM)* 10 (4): 10–59.

Cerf, Vinton G. 1989. "Requiem for the ARPANET." *ConneXions*, October: 27.

Cerf, Vinton G. 1990. Interview by Judy O'Neill, Reston, Virginia, 24 April. Charles Babbage Institute.

Cerf, Vinton G. 1993. "How the Internet Came to Be." In *The Online User's Encyclopedia*, ed. B. Aboba. Addison-Wesley.

Cerf, Vinton G., and Robert E. Kahn. 1974. "A Protocol for Packet Network Intercommunication." *IEEE Transactions on Communications* COM-22 (May): 637–648.

Cerf, Vinton G., and Peter T. Kirstein. 1978. "Issues in Packet-Network Interconnection." *Proceedings of the IEEE* 66 (11): 1386–1408.

Cerf, Vinton G., and Robert E. Lyons. 1983. "Military Requirements for Packet-Switched Networks and Their Implications for Protocol Standardization." *Computer Networks* 7: 293–306.

Cerf, Vinton G., Alex McKenzie, Roger Scantlebury, and Hubert Zimmermann. 1976. "Proposal for an International End to End Protocol." *ACM Computer Communication Review* 6 (1): 63.

Clark, David D. 1982. Names, Addresses, Ports, and Routes. RFC 814.

Clark, Wesley. 1990. Interview by Judy O'Neill, New York, 3 May. Charles Babbage Institute.

Cohen, Danny. 1978. "On Interconnection of Computer Networks." In Proceedings of Interlinking of Computer Networks, Bonas, France.

Comerford, Richard. 1995. "The Web: A Two-Minute Tutorial." *IEEE Spectrum* 32: 71.

Constant, Edward W. 1980. *The Origins of the Turbojet Revolution.* Johns Hopkins University Press.

Coopey, Richard. 1993. "Industrial Policy in the White Heat of the Scientific Revolution." In *The Wilson Governments, 1964–1970,* ed. R. Coopey et al. St. Martin's Press.

Coopey, Richard, and Donald Clarke. 1995. *3i : Fifty Years Investing in Industry.* Oxford University Press.

Cornew, Ronald W., and Philip M. Morse. 1975. "Distributive Computer Networking: Making It Work on a Regional Basis." *Science* 189 (August): 523–531.

Crispin, Mark. 1991. Message to comp.protocols.tcp-ip, 21 June. McKenzie box 2, Bolt, Beranek and Newman library.

Crocker, David. 1993. "Making Standards the IETF Way." *Standard View* 1 (1): 48–54.

Crocker, Stephen. 1969. Documentation Conventions. RFC 3.

Crocker, Stephen, John Heafner, Robert Metcalfe, and Jonathan Postel. 1972. "Function-Oriented Protocols for the ARPA Computer Network." In *Proceedings of AFIPS Spring Joint Computer Conference.* AFIPS Press.

Curran, Alex, and Vinton Cerf. 1975. "The Science of Computer Communications and the CCITT." In *Proceedings of International Conference on Communications.* IEEE.

Danet, A., R. Despres, A. LeRest, G. Pichon, and S. Ritzenthaler. 1976. "The French Public Packet Switching Service: The Transpac Network." In *Proceedings of Third International Conference on Computer Communication.* North-Holland.

Data Processing. 1969. "Low-Cost Automatic Message Switching." *Data Processing,* November-December: 580–581.

David, Paul A., and Julie Ann Bunn. 1988. "The Economics of Gateway Technologies and Network Evolution: Lessons from Electricity Supply History." *Information Economics and Policy* 3: 165–202.

Davies, B. H., and A. S. Bates. 1982. "Internetworking in the Military Environment." In *Proceedings of IEEE Computer Society International Conference.* IEEE.

Davies, Donald W. 1965. Proposal for the Development of a National Communication Service for On-Line Data Processing. Memorandum, National Physical Laboratory, 15 December. In National Archive for the History of Computing.

Davies, Donald W. 1966a. A Computer Network for NPL. Memorandum, National Physical Laboratory, 28 July. In National Archive for the History of Computing.

Davies, Donald W. 1966b. Proposal for a Digital Communication Network. Memorandum, National Physical Laboratory, June. In National Archive for the History of Computing.

Davies, Donald W. 1968a. Communication Requirements for Real-time Systems. Memorandum, 12 January. In National Archive for the History of Computing.

Davies, Donald W. 1968b. Report on a Visit to the United States of America 20 April–10 May. Memorandum. In National Archive for the History of Computing.

Davies, Donald W. 1968c. Communication Requirements of a Mintech Computer Network. Memorandum, National Physical Laboratory, November. In National Archive for the History of Computing.

Davies, Donald W. 1986. Interview by Martin Campbell-Kelly, 17 March. In National Archive for the History of Computing.

Day, John. 1997. Interview by Janet Abbate, Cambridge, Massachusetts, 13 May.

Department of Defense. 1972. "Project THEMIS." In *The Politics of American Science, 1939 to the Present,* ed. J. Penick Jr. et al. MIT Press.

Dhas, C. R., and V. K. Konangi. 1986. "X.25: An Interface to Public Packet Networks." *IEEE Communications Magazine* 24 (9): 18–25.

Dickson, Paul A. 1968. "ARPA Network Will Represent Integration on a Large Scale." *Electronics,* 30 September: 131–134.

Dorin, Robert H., and Donald E. Eastlake III. 1976. "Use of the Datacomputer in the VELA Seismological Network." In *Proceedings of IEEE Computer Society International Conference.* IEEE.

Edgerton, David. 1996. "The 'White Heat' Revisited: The British Government and Technology in the 1960s." *20th Century British History* 7 (1): 53–90.

Edwards, Paul N. 1996. *The Closed World: Computers and the Politics of Discourse in Cold War America.* MIT Press.

Electronics. 1972. "Demonstration Heralds Next Wave: Connecting a Network of Networks." *Electronics,* 6 November: 34–36.

Electronics. 1978. "Protocol-Linkup Plan Would Stymie IBM." *Electronics,* 8 June: 69–70.

Emery, James C. 1976. "Development of a Facilitating Network for Higher Education." In *Proceedings of IEEE Computer Society International Conference (Compcon).* IEEE.

Farber, David J. 1972. "Networks: An Introduction." *Datamation,* April: 36–39.

Feinler, Jake (Elizabeth). 1982a. *ARPANET Newsletter* 13 (12 July).

Feinler, Jake (Elizabeth). 1982b. *ARPANET Newsletter* 16 (30 September).

Figallo, Cliff. 1995. "The WELL: A Regionally Based On-Line Community on the Internet." In *Public Access to the Internet,* ed. B. Kahin and J. Keller. MIT Press.

Flamm, Kenneth. 1988. *Creating the Computer: Government, Industry, and High Technology.* Brookings Institution.

Folts, Harold C. 1978. "Interface Standards for Public Data Networks." In *Proceedings of IEEE Computer Society International Conference.* IEEE.

Foster, John S. 1972. "Hearings on National Science Policy, H. Con. Res. 666." In *The Politics of American Science, 1939 to the Present,* ed. J. Penick Jr. et al. MIT Press.

Foy, Nancy. 1986. "Britain's Real-Time Club." *Annals of the History of Computing* 8 (4): 370–371.

Frank, Howard. 1990. Interview by Judy O'Neill. Fairfax, Virginia, 30 March. Charles Babbage Institute.

Frank, Howard, I. T. Frisch, and W. Chou. 1970. "Topological Considerations in the Design of the ARPA Computer Network." In *Proceedings of AFIPS Spring Joint Computer Conference.* AFIPS Press.

Frank, Howard, Robert E. Kahn, and Leonard Kleinrock. 1972. "Computer Communication Network Design—Experience with Theory and Practice." In *Proceedings of AFIPS Spring Joint Computer Conference.* AFIPS Press.

Goldberg, Adele, ed. 1988. *A History of Personal Workstations.* ACM Press.

Goodfellow, Geoff. 1997. Before the DNS: How I upstaged the NIC's Official HOSTS.TXT. Message to Community Memory mailing list, 15 November. Archived at http: //memex.org/community-memory.

Gorgas, J. W. 1968. "The Polygrid Network for AUTOVON." *Bell Laboratories Record* 46 (4): 223–227.

Greenberger, Martin, Julius Aronofsky, James L. McKenney, and William F. Massy. 1973. "Computer and Information Networks." *Science* 182: 29–35.

Hadley, E. E., and B. R. Sexton. 1978. "The British Post Office Experimental Packet Switched Service (EPSS)—A Review of Development and Operational Experience." In *Proceedings of Fourth International Conference on Computer Communication.* North-Holland.

Hafner, Katie, and Matthew Lyon. 1996. *Where Wizards Stay Up Late: The Origins of the Internet.* Simon & Schuster.

Harris, Thomas C., Peter V. Abene, Wayne W. Grindle, Darryl W. Henry, Dennis C. Morris, Glynn E. Parker, and Jeffrey Mayersohn. 1982. "Development of the MILNET." In *Proceedings of IEEE Electronics and Aerospace Systems Conference (EASCON)*. IEEE.

Haughney, Major Joseph. 1980a. *ARPANET Newsletter* 1 (1 July).

Haughney, Major Joseph. 1980b. *ARPANET Newsletter* 2 (August).

Haughney, Major Joseph. 1980c. *ARPANET Newsletter* 3 (10 September).

Haughney, Major Joseph. 1981a. *ARPANET Newsletter* 6 (30 March).

Haughney, Major Joseph. 1981b. *ARPANET Newsletter* 8 (15 September).

Hayes, Brian. 1994. "The World Wide Web." *American Scientist* 82: 416–420.

Heart, Frank. 1990. Interview by Judy O'Neill, Cambridge, Massachusetts, 13 March. Charles Babbage Institute.

Heart, Frank, Robert Kahn, Severo Ornstein, William Crowther, and David Walden. 1970. "The Interface Message Processor for the ARPA Computer Network." AFIPS Spring Joint Computer Conference.

Heart, Frank, Alex McKenzie, John McQuillan, and David Walden. 1977. Draft ARPANET Completion Report, 9 September (bound computer printout, Bolt, Beranek and Newman). McKenzie box 2, Bolt, Beranek and Newman library.

Heart, Frank, Alex McKenzie, John McQuillan, and David Walden. 1978. *ARPANET Completion Report*, 4 January. McKenzie box 2, Bolt, Beranek and Newman library.

Heiden, Heidi B. 1982. *ARPANET Newsletter* 19 (2 December).

Heiden, Heidi B. 1983a. *ARPANET Newsletter* 20 (13 January).

Heiden, Heidi B. 1983b. *ARPANET Newsletter* 23 (7 April).

Heiden, Heidi B., and Howard C. Duffield. 1982. "Defense Data Network." In *Proceedings of IEEE Electronics and Aerospace Systems Conference (EASCON)*. IEEE.

Heilmeier, George H. 1976. "The Role of DARPA." *Commanders Digest*, 7 October: 2–8.

Hendry, John. 1990. *Innovating for Failure: Government Policy and the Early British Computer Industry*. MIT Press.

Henriques, Vico E. 1975. "The American National Standards Committee X3: Computers and Information Processing." In *Proceedings of IEEE Computer Society International Conference*. IEEE.

Hirsch, Phil. 1975. "Canada Network Won't Take SDLC Protocol." *Datamation*, March: 121–123.

Hirsch, Phil. 1976a. "Protocol Control: Carriers or Users?" *Datamation,* March: 188–189.

Hirsch, Phil. 1976b. "Protocol for Packet Networks: The Question in Implementation." *Datamation,* May: 187–190.

Horner, David. 1993. "The Road to Scarborough: Wilson, Labour and the Scientific Revolution." In *The Wilson Governments, 1964–1970,* ed. R. Coopey et al. St. Martin's.

Hornig, Charles. 1984. A Standard for the Transmission of IP Datagrams over Ethernet Networks. RFC 894.

Hughes, Thomas P. 1998. *Rescuing Prometheus.* Pantheon.

IEEE Spectrum. 1964. "Scanning the Issues." *IEEE Spectrum,* August: 114.

IFIP Working Group 6.1. 1979. "Implications of Recommendation X.75 and Proposed Improvements for Public Data Network Interconnection." *Computer Communication Review (ACM)* 9 (3): 33–39.

ISO (International Organization for Standardization). 1984a. ISO Transport Protocol Specification: ISO DP 8073. RFC 905.

ISO. 1984b. Protocol for Providing the Connectionless-Mode Network Services (Informally—ISO IP). RFC 926.

ISO Technical Committee 97, Subcommittee 16. 1978. "Provisional Model of Open-Systems Architecture." *Computer Communication Review (ACM)* 8 (3): 49–62.

Jacobs, Irwin Mark, Richard Binder, and Estil V. Hoversten. 1978. "General Purpose Packet Satellite Networks." *Proceedings of the IEEE* 66 (11): 1448–1467.

Jennings, Dennis M., Lawrence H. Landweber, Ira H. Fuchs, David J. Farber, and W. Richards Adrion. 1986. "Computer Networking for Scientists." *Science* 231 (4741): 943–950.

Johnson, Lyndon B. 1972. "Statement of the President to the Cabinet on Strengthening the Academic Capability for Science throughout the Nation." In *The Politics of American Science, 1939 to the Present,* ed. J. Penick Jr. et al. MIT Press.

Kahn, Robert E. 1975. "The Organization of Computer Resources into a Packet Radio Network." In *Proceedings of AFIPS National Computer Conference.* AFIPS Press.

Kahn, Robert E. 1989. Interview by William Aspray, Reston, Virginia, 22 March. Charles Babbage Institute.

Kahn, Robert E. 1990. Interview by Judy O'Neill, Reston, Virginia, 24 April. Charles Babbage Institute.

Kahn, Robert E., Steven A. Gronemeyer, Jerry Burchfiel, and Ronald C. Kunzelman. 1978. "Advances in Packet Radio Technology." *Proceedings of the IEEE* 66 (11): 1468–1496.

Karp, Peggy M. 1973. "Origin, Development and Current Status of the ARPA Network." In *Proceedings of IEEE Computer Society International Conference.* IEEE.

Kellerman, Aharon. 1993. *Telecommunications and Geography.* Belhaven.

Kelly, P. T. F. 1978. "Public Packet Switched Data Networks, International Plans and Standards." *Proceedings of the IEEE* 66 (11): 1539–1549.

Kerr, Anne V., ed. 1985. *The VELA Program: A 25-Year Review of Basic Research.* DARPA.

Kirstein, P. T. 1976. "Management Questions in Relationship to the University College London Node of the ARPA Computer Network." In *Proceedings of Third International Conference on Computer Communication.* North-Holland.

Klass, Philip J. 1976. "ARPA Net Aids Command, Control Tests." *Aviation Week & Space Technology,* 27 September: 63–67.

Kleinrock, Leonard. 1976. *Queuing Systems.* Wiley.

Kleinrock, Leonard. 1990. Interview by Judy O'Neill, Los Angeles, 3 April. Charles Babbage Institute.

Krol, Ed. 1992. *The Whole Internet User's Guide & Catalog.* O'Reilly & Associates.

Kubrick, Stanley, director. 1963. *Dr. Strangelove, or How I Learned to Stop Worrying and Love the Bomb.* Screenplay by Stanley Kubrick, Peter George, and Terry Southern, based on the book *Red Alert* by Peter George. Copyright Hawk Films Ltd. Video distributed by RCA/Columbia Pictures Home Video.

Kunzelman, Ronald C. 1978. "Overview of the ARPA Packet Radio Experimental Network." In *Proceedings of IEEE Computer Society International Conference.* IEEE.

Kuo, Franklin F. 1974. "Political and Economic Issues for Internetwork Connections." In *Proceedings of Second International Conference on Computer Communication.* North-Holland.

Kuo, Franklin F. 1975. "Public Policy Issues Concerning ARPANET." In *Proceedings of IEEE Fourth Data Communications Symposium.* IEEE.

Kuo, Franklin F. 1978. "Defense Packet Switching Networks in the United States." In Proceedings of Interlinking of Computer Networks, Bonas, France.

Lamond, Fred. 1985. "IBM's Standard Stand." *Datamation,* 1 February: 106.

Landweber, Lawrence H. 1991. CSNET: A Brief History. Message to com-priv and ietf newsgroups, 22 September.

Landweber, Lawrence H., and Marvin H. Solomon. 1982. "Use of Multiple Networks in CSNET." In *Proceedings of IEEE Computer Society International Conference*. IEEE.

Latham, Donald D. 1987. Open Systems Interconnection Protocols. Memorandum for Secretaries of the Military Departments; Chairman, Joint Chiefs of Staff; Directors, Defense Agencies, 2 July.

Laws, J., and V. Hathway. 1978. "Experience from Two Forms of Inter-Network Connection." In Proceedings of Interlinking of Computer Networks, Bonas, France.

Lederberg, Joshua. 1978. "Digital Communications and the Conduct of Science: The New Literacy." *Proceedings of the IEEE* 66 (11): 1314–1319.

Lederberg, Joshua. 1997. Message to Community Memory mailing list, 26 March. Archived at http: //memex.org/community-memory.

Leiner, Barry M., Vinton G. Cerf, David D. Clark, Robert E. Kahn, Leonard Kleinrock, Daniel C. Lynch, Jon Postel, Larry G. Roberts, and Stephen Wolff. 1997. "A Brief History of the Internet." Web page http://www.isoc.org/internet-history.

Licklider, J. C. R. 1960. "Man-Computer Symbiosis." *IRE Transactions on Human Factors in Electronics* 1 (1): 4–11.

Licklider, J. C. R., and Albert Vezza. 1978. "Applications of Information Networks." *Proceedings of the IEEE* 66: 1330–1345.

Long Lines (AT&T magazine). 1965. "[AUTOVON]." 45: 3. 1–7.

Long Lines. 1969. "[AUTOVON]." 48: 9. 18–23.

Lukasik, Stephen. 1973. "ARPA." *Commanders Digest,* 9 August: 2–12.

Lukasik, Stephen. 1991. Interview by Judy O'Neill, Redondo Beach, California, 17 October. Charles Babbage Institute.

Lycos, Peter, ed. 1975. *Computer Networking and Chemistry*. American Chemical Society.

Lynch, Dan. 1991. Message to comp.protocols.tcp-ip, 23 June. McKenzie box 2, Bolt, Beranek and Newman library.

Lynch, Daniel C., and Marshall T. Rose, eds. 1993. *Internet System Handbook*. Addison-Wesley.

MacDonald, V. C. 1978. "Domestic and International Standards Activities for Data Communications." In *Proceedings of IEEE Computer Society International Conference*. IEEE.

MacKenzie, Donald. 1991. "The Influence of the Los Alamos and Livermore National Laboratories on the Development of Supercomputing." *Annals of the History of Computing* 13: 179–200.

Malik, Rex. 1989. "The Real Time Club." *Annals of the History of Computing* 11 (1): 51–52.

Marill, Thomas, and Lawrence G. Roberts. 1966. "Toward a Cooperative Network of Time-Shared Computers." In *Proceedings of AFIPS Fall Joint Computer Conference.* AFIPS Press.

Massy, William F. 1974. "Computer Networks: Making the Decision to Join One." *Science* 186: 414–420.

McGovern, P. J. 1973. *EDP Industry Report, 6 December.* International Data Corporation.

McKenzie, Alex. 1976. "The ARPA Network Control Center." In *Computer Networks,* ed. M. Abrams et al. IEEE.

McKenzie, Alex. 1990. Interview by Judy O'Neill, Cambridge, Massachusetts, 13 March. Charles Babbage Institute.

McKenzie, Alex. 1991a. Email to Michael Hennebe, 20 May. McKenzie box 2, Bolt, Beranek and Newman library.

McKenzie, Alex. 1991b. Message to comp.protocols.tcp-ip, 24 June. McKenzie box 2, Bolt, Beranek and Newman library.

McKenzie, Alex. 1997. Interview by Janet Abbate, Lincoln, Massachusetts, 12 May.

McQuillan, John M., and David C. Walden. 1977. "The ARPA Network Design Decisions." *Computer Networks* 1: 243–289.

McQuillan, J. M., W. R. Crowther, B. P. Cosell, D. C. Walden, and F. E. Heart. 1972. "Improvements in the Design and Performance of the ARPA Network." In Proceedings of AFIPS Fall Joint Computer Conference.

McWilliams, Gary. 1987. "They Just Can't Wait to Integrate." *Datamation,* 15 February: 17–18.

MERIT. 1989. *[According to the copy, this is the source of figure 6.1—please insert the reference in proofs]*

MERIT. 1995a. "Merit Retires NSFNET Backbone Service." nic.merit.edu/nsfnet/news.releases/nsfnet.retired.

MERIT. 1995b. "nets.by.country." 1 May. http://nic.merit.edu/nsfnet/statistics/nets.by.country.

MERIT. 1997. "history.hosts." 1 September. http://nic.merit.edu/nsfnet/statistics/history.hosts.

Metcalfe, Robert M. 1996. "Packet Communication." In *Computer Classics Revisited,* ed. P. Salus. Peer-to-Peer Communications.

Mills, David L. 1981. Internet Name Domains. RFC 799.

Morgan, Austen. 1992. *Harold Wilson*. Pluto.

Negroponte, Nicholas. 1995. *Being Digital*. Knopf.

Neil, Suzanne, Lee McKnight, and Joseph Bailey. 1995. "The Government's Role in the HDTV Standards Process: Model or Aberration?" In *Standards Policy for Information Infrastructure*, ed. B. Kahin and J. Abbate. MIT Press.

Norberg, Arthur L., and Judy E. O'Neill. 1992. *A History of the Information Processing Techniques Office of the Defense Advanced Research Projects Agency*. Charles Babbage Institute.

Norberg, Arthur L., and Judy E. O'Neill. 1996. *Transforming Computer Technology: Information Processing for the Pentagon, 1962–1986*. Johns Hopkins University Press.

NSF Network Technical Advisory Group. 1986. Requirements for Internet Gateways—Draft. RFC 985.

O'Neill, E. F., ed. 1985. *A History of Engineering and Science in the Bell System: Transmission Technology (1925–1975)*. AT&T Bell Laboratories.

Ornstein, S. M., F. E. Heart, W. R. Crowther, H. K. Rising, S. B. Russell, and A. Michel. 1972. "The Terminal IMP for the ARPA Computer Network." AFIPS Spring Joint Computer Conference.

Padlipsky, M. A. 1983. "A Perspective on the ARPANET Reference Model." In Proceedings of IEEE INFOCOM 1983.

Parker, Major Glynn. 1982a. *ARPANET Newsletter* 11 (2 March).

Parker, Major Glynn. 1982b. *ARPANET Newsletter* 12 (11 June).

Parker, Major Glynn, and Vinton Cerf. 1982. *ARPANET Newsletter* 10 (12 February).

Passmore, L. David. 1985. "The Networking Standards Collision." *Datamation*, 1 February: 98–108.

Perillo, Francine. 1981. *ARPANET Newsletter* 9 (21 October).

Pollack, Andrew. 1989. "America's Answer to Japan's MITI." *New York Times*, 5 May.

Postel, Jon. 1982. Simple Mail Transfer Protocol. RFC 821.

Postel, Jon, and J. Reynolds. 1984. ARPA-Internet Protocol Policy. RFC 902.

Postel, Jon, Carl Sunshine, and D. Cohen. 1982. "Recent Developments in the DARPA Internet Program." In Proceedings of Sixth International Conference on Computer Communication.

Pouzin, Louis. 1975a. "The Communications Network Snarl." *Datamation*, December: 70–72.

Pouzin, Louis. 1975b (reprint of 1973 paper). "Presentation and Major Design Aspects of the CYCLADES Computer Network." In *Computer Communication Networks,* ed. R. Grimsdale and F. Kuo. Noordhoff.

Pouzin, Louis, and Hubert Zimmermann. 1978. "A Tutorial on Protocols." *Proceedings of the IEEE* 66 (11): 1386–1408.

Pritchard, Wilbur L. 1984. "The History and Future of Commercial Satellite Communications." *IEEE Communications Magazine* 22 (5): 22–37.

PSINet. 1997. "PSINet: Our History." Web page http://www.psi.net/profile/history.html.

Pyatt, Edward. 1983. *The National Physical Laboratory: A History.* Hilger.

Quarterman, John S. 1990. *The Matrix: Computer Networks and Conferencing Systems Worldwide.* Digital Press.

Quarterman, John S., and Josiah C. Hoskins. 1986. "Notable Computer Networks." *Communications of the ACM* 29: 932–971.

RCA Service Company, Government Services Division. 1972. ARPANET Study Final Report. Contract no. F08606–73-C-0018.

Retz, David L., and Bruce W. Schafer. 1976. "Structure of the ELF Operating System." In *Proceedings of AFIPS National Computer Conference.* AFIPS Press.

Reynolds, J., and J. Postel. 1987. The Request For Comments Reference Guide. RFC 1000.

Rinde, J. 1976. "TYMNET I: An Alternative to Packet Technology." In *Proceedings of Third International Conference on Computer Communication.* North-Holland.

RIPE. 1997. "About RIPE". Web page http://www.ripe.net/info/ripe/ripe.html.

Roberts, Lawrence G. 1967a. Message Switching Network Proposal. ARPA document, spring 1967. National Archives Branch Depository, Suitland, Maryland, RG 330–78-0085, box 2, folder "Networking 1968–1972."

Roberts, Lawrence G. 1967b. "Multiple Computer Networks and Intercomputer Communication." In Proceedings of ACM Symposium on Operating System Principles, Gatlinburg, Tennessee.

Roberts, Lawrence G. 1970. "Communications and System Architecture." In *Proceedings of IEEE Computer Society International Conference.* IEEE.

Roberts, Lawrence G. 1973. "Network Rationale: A 5-Year Re-evaluation." In *Proceedings of IEEE Computer Society International Conference.* IEEE.

Roberts, Lawrence G. 1974. "Data by the Packet." *IEEE Spectrum,* February: 46–51.

Roberts, Lawrence G. 1978. "The Evolution of Packet Switching." *Proceedings of the IEEE* 66 (11): 1307–1313.

Roberts, Lawrence G. 1988. "The ARPANET and Computer Networks." In *A History of Personal Workstations,* ed. A. Goldberg. ACM Press.

Roberts, Lawrence G. 1989. Interview by Arthur L. Norberg, San Mateo, California, 4 April. Charles Babbage Institute.

Roberts, Lawrence G., and Robert E. Kahn. 1972. "Special Project: Participating Demonstration of a Multi-Purpose Network Linking Dissimilar Computers and Terminals." In *Proceedings of International Conference on Computer Communication.* North-Holland.

Roberts, Lawrence G., and Barry D. Wessler. 1970. "Computer Network Development to Achieve Resource Sharing." In *Proceedings of AFIPS Spring Joint Computer Conference.* AFIPS Press.

Rochlin, Gene. 1997. *Trapped in the Net: The Unanticipated Consequences of Computerization.* Princeton University Press.

Rulifson, Jeff. 1969. DEL. RFC 5.

Rybczynski, A., B. Wessler, R. Despres, and J. Wedlake. 1976. "A New Communication Protocol for Accessing Data Networks: The International Packet-Mode Interface." In *Proceedings of IEEE International Conference on Communications.* IEEE.

Salus, Peter H. 1995. *Casting the Net: From ARPANET to Internet and Beyond.* Addison-Wesley.

Sax, John. 1991. Email to Alex McKenzie, 20 May. McKenzie box 2, Bolt, Beranek and Newman library.

Schatz, Bruce R., and Joseph B. Hardin. 1994. "NCSA Mosaic and the World Wide Web: Global Hypermedia Protocols for the Internet." *Science* 265: 895–901.

Schelonka, Edward P. 1976. "Resource Sharing with Arpanet." In *Computer Networks,* ed. M. Abrams et al. IEEE.

Schindler, G. E., ed. 1982. *A History of Engineering and Science in the Bell System: Switching Technology (1925–1975).* AT&T Bell Laboratories.

Schultz, Brad. 1988. "The Evolution of ARPANET." *Datamation,* August: 71–74.

Schutz, Gerald C., and George E. Clark. 1974. "Data Communication Standards." *IEEE Computer* 7 (February): 32–37.

Shapard, Jeffrey. 1995. "Islands in the (Data)Stream: Language, Character Codes, and Electronic Isolation in Japan." In *Global Networks,* ed. L. Harasim. Sage.

Shapero, Albert, Richard P. Howell, and James R. Tombaugh. 1965. *The Structure and Dynamics of the Defense R&D Industry: The Los Angeles and Boston Complexes.* Stanford Research Institute.

Sher, Michael S. 1974. "A Case Study in Networking." *Datamation,* March: 56–59.

Shute, Nevil. 1957. *On the Beach.* Morrow.

Smith, Ben. 1995. "Internet with Style." *Byte:* 197–200.

Sproull, Lee, and Sara Kiesler. 1991. *Connections: New Ways of Working in the Networked Organization.* MIT Press.

Stallings, William. 1991. *Data and Computer Communications,* third edition. Macmillan.

Sunshine, Carl A. 1981. "Transport Protocols for Computer Networks." In *Protocols and Techniques for Data Communication Networks,* ed. F. Kuo. Prentice-Hall.

Tanenbaum, Andrew S. 1989. *Computer Networks,* second edition. Prentice-Hall.

Taylor, Robert. 1989. Interview by William Aspray, Palo Alto, California, 28 February. Charles Babbage Institute.

Thacker, Chuck. 1988. "Personal Distributed Computing: The Alto and Ethernet Hardware." In *A History of Personal Workstations,* ed. A. Goldberg. ACM Press.

Travers, Ginny. 1991. Email to Alex McKenzie, 22 May. McKenzie box 2, Bolt, Beranek and Newman library.

Turkle, Sherry. 1995. *Life on the Screen: Identity in the Age of the Internet.* Simon & Schuster.

Turoff, Murray, and Starr Roxanne Hiltz. 1977. "Meeting through Your Computer." *IEEE Spectrum* 14: 58–64.

Twyver, D. A., and A. M. Rybczynski. 1976. "Datapac Subscriber Interfaces." In *Proceedings of Third International Conference on Computer Communication.* North-Holland.

US Congress (House of Representatives). 1967. Subcommittee on Appropriations. Hearings on Department of Defense Appropriations for Fiscal Year 1968. Testimony by Dr. John S. Foster, Director of Defense Research and Engineering. 90th Congress, first session, 3: 138, 184–186.

US Congress (House of Representatives). 1969. Hearings on Department of Defense Appropriations for Fiscal Year 1970. 91st Congress, first session, 809.

US Congress (House of Representatives). 1971. Subcommittee on Appropriations. Hearings on Department of Defense Appropriations for Fiscal Year

1972. Testimony by Stephen Lukasik, Director of ARPA. 92nd Congress, first session, 7: 282–284, 287, 324–326.

US Congress (House of Representatives). 1974. Subcommittee on Appropriations. Hearings on Department of Defense Appropriations for Fiscal Year 1975. Director of ARPA Stephen Lukasik. 93rd Congress, second session, 36–40 (of included text).

US Congress (Senate). 1968. Subcommittee on Appropriations. Hearings on Department of Defense Appropriations for Fiscal Year 1968. Director of Defense Research and Engineering Dr. John S. Foster. 90th Congress, first session, 2248–2249, 2305–2308, 2346–2348.

US Congress (Senate). 1969. Subcommittee on Appropriations. Hearings on Department of Defense Appropriations for Fiscal Year 1970. Testimony by Eberhardt Rechtin, Director of ARPA. 91st Congress, first session, 441–442.

US Congress (Senate). 1971. Subcommittee on Appropriations. Hearings on Department of Defense Appropriations for Fiscal Year 1972. Lukasik. 92nd Congress, first session, 1: 646–647, 652, 696, 736–738.

US Congress (Senate). 1972. Subcommittee on Appropriations. Hearings on Department of Defense Appropriations for Fiscal Year 1973. Director of ARPA Stephen Lukasik. 92nd Congress, second session, 745, 821–823.

US Congress (Senate). 1991. High-Performance Computing Act of 1991 (S272). Sponsored by Albert Gore. 102nd Congress, first session.

UCLA Campus Computing Network. 1974. "The ARPA Computer Network (ARPANET)." *CCN Newsletter,* June 24.

von Alven, William H. 1974. "The Modem—A Passing Fancy?" In *Proceedings of IEEE Computer Society International Conference.* IEEE.

Watson, Richard W. 1971. A Mail Box Protocol. RFC 196.

Werbach, Kevin. 1997. Digital Tornado: The Internet and Telecommunications Policy. Working Paper 29, Office of Plans and Policy, Federal Communications Commission.

Wheeler, John L. 1975. "International Data Communication Standards." In *Proceedings of IEEE Computer Society International Conference.* IEEE.

Wichmann, B. A. 1967. An Electronic Typewriter. Memorandum, National Physical Laboratory, 26 May. In National Archive for the History of Computing.

Wilkes, M. V. 1980. "Computers into the 1980s." *Electronics & Power,* January: 69–70.

Wilkinson, P. T. 1968. The Main Features of a Proposed National Data Communication Network. Memorandum, National Physical Laboratory, February. In National Archive for the History of Computing.

Wilson, Harold. 1971. *The Labour Government, 1964–1970: A Personal Record.* Weidenfeld and Nicolson.

Wolff, Stephen. 1991. Merit Retires NSFNET Backbone Service. Email to com-priv and farnet members, 26 November. See nic.merit.edu/cise/pdp.txt.

Wood, B. M. 1982. "Open Systems Interconnection—Basic Concepts and Current Status." In *Proceedings of Sixth International Conference on Computer Communication.* North-Holland.

Zimmermann, Hubert. 1976. "High Level Protocols Standardization: Technical and Political Issues." In *Proceedings of Third International Conference on Computer Communication.* North-Holland.

Zraket, Charles A. 1990. Interview by Arthur L. Norberg, Bedford, Massachusetts, 3 May. Charles Babbage Institute.

Index